Merging
Languages and Engineering

Partnering Across the Disciplines

Synthesis Lectures on Global Engineering

Editor
Gary Downey, *Virginia Tech*
Assistant Editor
Kacey Beddoes, *Purdue*

The Global Engineering Series challenges students, faculty and administrators, and working engineers to cross the borders of countries, and it follows those who do. Engineers and engineering have grown up within countries. The visions engineers have had of themselves, their knowledge, and their service have varied dramatically over time and across territorial spaces. Engineers now follow diasporas of industrial corporations, NGOs, and other transnational employers across the planet. To what extent do engineers carry their countries with them? What are key sites of encounters among engineers and non-engineers across the borders of countries? What is at stake when engineers encounter others who understand their knowledge, objectives, work, and identities differently? What is engineering now for? What are engineers now for?

The Series invites short manuscripts making visible the experiences of engineers and engineering students and faculty across the borders of countries. Possible topics include engineers in and out of countries, physical mobility and travel, virtual mobility and travel, geo-spatial distributions of work, international education, international work environments, transnational identities and identity issues, transnational organizations, research collaborations, global normativities, and encounters among engineers and non-engineers across country borders.

The Series juxtaposes contributions from distinct disciplinary, analytical, and geographical perspectives to encourage readers to look beyond familiar intellectual and geographical boundaries for insight and guidance. Holding paramount the goal of high-quality scholarship, it offers learning resources to engineering students and faculty and working engineers crossing the borders of countries. Its commitment is to help them improve engineering work through critical self-analysis and listening.

Merging Languages and Engineering: Partnering Across the Disciplines
John M. Grandin
2013

Merging Languages and Engineering: Partnering Across the Disciplines
John M. Grandin

ISBN: 978-3-031-00999-0   paperback
ISBN: 978-3-031-02127-5   ebook

DOI: 10.1007/978-3-031-02127-5

A Publication in the Springer series
SYNTHESIS LECTURES ON GLOBAL ENGINEERING

Lecture #3
Series Editor: Gary Downey, *Virginia Tech*
Assistant Editor: Kacey Beddoes, *Purdue*
Series ISSN
Synthesis Lectures on Global Engineering
Print 2160-7664   Electronic 2160-7672

Merging Languages and Engineering

Partnering Across the Disciplines

John M. Grandin
University of Rhode Island

SYNTHESIS LECTURES ON GLOBAL ENGINEERING #3

ABSTRACT

At the University of Rhode Island over 25% of engineering undergraduates simultaneously complete a second degree in German, French, Spanish, or Chinese. They furthermore spend an entire year abroad, one semester as exchange students at a partner university and six months as professional engineering interns at a cooperating company. With a close-to 100% placement rate, over 400 graduates, and numerous national awards, the URI International Engineering Program (IEP) is a proven path of preparation for young engineers in today's global workplace.

The author of this volume, John Grandin, is an emeritus professor of German who developed and led the IEP for twenty-three years. In these pages, he provides a two-pronged approach to explain the origin and history of this program rooted in such an unusual merger of two traditionally distinct higher education disciplines. He looks first at himself to explain how and why he became an international educator and what led him to his lasting passion for the IEP. He then provides an historical overview of the program's origin and growth, including looks at the bumps and bruises and ups and downs along the way. Grandin hopes that this story will be of use and value to other educators determined to reform higher education and align it with the needs of the 21st Century.

KEYWORDS

languages and engineering, international engineering, global engineering education, internationalization of the curriculum, international education, cross-disciplinary education, crossing the disciplines, partnering across the disciplines, educating for the global workplace

Dedicated to
the students and graduates of the IEP.
They are my heroes.

Contents

Preface

On the surface it does not seem likely that engineering undergraduates would want to simultaneously enroll in foreign language classes. But they do at the University of Rhode Island. Over one-quarter of the URI engineering students not only study German, French, Spanish, or Mandarin Chinese, but complete a full major in one of those languages and spend a year abroad as exchange students and professional interns, thereby completing both the BA with a language major and the BS degree in their chosen engineering field as the major components of URI's now well-known International Engineering Program (IEP).

The IEP has over 400 graduates in the workplace today, enrolls over 300 students, and sends up to 60 students abroad each year for study and internships with partner companies. With close to a 100% job placement rate, it has proven itself as a curriculum appropriate for our age of globalization. The workplace is international, and business and industry direly need young engineers who are mobile, flexible, multilingual, and capable of working efficiently in global teams, whether from a home base or at any location around the world.

As the recently retired director and main developer of the IEP, I am often asked how the program came about. When did it start? Who was behind it? Why did it happen at URI? How did it happen? What would one need to build such a program elsewhere? And, as one of the most common questions, how did I, not as an engineer, but as a professor of German, get involved in such an undertaking? What made me want to do this?

In the following pages, therefore, I will attempt to answer these and other questions in two ways. Autobiographically, I will pursue my own path to German language teaching, to international education, and ultimately to the IEP. Though this may cause you to learn more about me than you want to know, I think it would be useful to explore what made me tick in this process and the extent to which such a curricular development hinges on individuals with an unusual commitment. Secondly, I will do my best to record the history of the IEP over my 23 years at the helm, both in terms of the easy parts and the frustratingly difficult parts. Though I may step on some toes in the process, I hope it will be of value for the IEP and for others who seek to innovate and improve higher education to read about the barriers encountered along the way and how we managed to get by them.

I would like to thank Professor Gary Downey from Virginia Tech who invited me to write the first version of this book as a chapter for his recently published volume on international engineering education.[1] Gary believed we could learn how such programs are created by not just describing the programs themselves, but by also exploring the "personal geographies" of the main forces behind

---

[1]Grandin, John, "Bridging Two Worlds," in Downey, Gary and Beddoes, Kacey, Editors, *What is Global Engineering Education For? The Making of International Educators*, Morgan & Claypool, 2011.

them. Gary liked my chapter enough to encourage me to write more, the result of which I share with you now.

John M. Grandin
Ocean Park, Maine, and Wakefield, Rhode Island
December 2012

CHAPTER 1

How I became a Professor of German

People sometimes ask if I am German or of German heritage, in search of my reason for becoming professor of German and international educator. But my path to that career was driven neither by ancestry nor even deliberate planning. It harks back first of all to advice from my clergyman father shortly before I was to start my freshman year at Kalamazoo College, a small liberal arts college in Michigan. Since he envisioned me following in his footsteps as a minister and knew I was interested in philosophy, he advised me that learning to read German would be sound preparation for studying the great philosophers and theologians. So I signed up for German 101, even though Latin in high school had been disastrous for me and the thought of another language was really not at all appealing. To my surprise, German turned out to be my favorite course during my freshman year, largely because I really admired the instructor, an older gentleman (from my perspective then) from the local high school who had been hired as a part-timer to teach my class. He motivated me to work hard and my strongest grades were in that class. Thus, I was happy to enroll in intermediate German for my sophomore year, still unaware of how decisive that step would be.

Unbeknownst to me, the Chairman of the Board of Trustees at Kalamazoo College had recently given the school 2.5 million dollars to create an endowment to support experiences abroad for "K College" students. Dr. Richard U. Light had recently taken his family to Grenoble, France, for a summer of intensive language courses, and was so moved by the overall cultural experience and the progress with the language made by his four sons, that he wanted to see this become a basic part of a liberal arts education. I was thus taken fully off guard when my German teacher, Frau Elizabeth Mayer, asked me after class one day if I wanted to spend the coming summer in Germany, living with a family in Bonn and taking an intensive German course at the university. In this unforeseen way, I became one of a group of 48 Kalamazoo College students dispersed for travel and study that summer in France, Spain, and Germany.

As the twenty-year-old son of a Baptist minister from a small, homogeneous community in central Massachusetts, being sent to Germany in the summer of 1960 ushered me into a very different world. I learned first that there were people who actually spoke German, and that I too could learn to communicate in that language and have fun doing it. Next I learned there were lots of very good people who had been shaped by very different experiences and who had insights into history, society, and life quite different from my own. My host "parents," the von Manteuffels (with whose children I still have contact), had lived through both world wars and had been driven out of their homeland in the formerly German East Prussia and forced to start a new life in the West. While their first motivation for taking me into their home that summer was no doubt their need for additional income to help support themselves and their four children, their curiosity about young people from America was a strong second reason. There was thus a lively exchange of views and opinions between us during this time, supported by a regularly consulted German-English dictionary on the middle of the dining room table ("Buch, Buch, Buch," we would say in desperate moments). Living in their home for three months, sitting with them at their dinner table every day, watching how they interacted as a family, hearing their concerns and aspirations, debating Kennedy versus Nixon, finding myself and my worldview challenged by them on a daily basis, the world took on new dimensions for me in that short summer, as it did for them as well. I found many of my own basic values and political assumptions challenged and began to understand, for example, that being an American was taken quite glibly and naively by my friends and myself and that most of us were totally unaware of the life experiences and priorities of people in other parts of the world, or of their expectations and hopes of us. Combining this with my travels on a $50 motorcycle through Germany, Belgium, Holland, Switzerland, Liechtenstein, and Austria, finding a German girlfriend, growing a beard for the first time, enjoying long discussions in youth hostels with students from numerous countries, as well as spending three days in East and West Berlin a year before the Wall was built, I went back home at the end of August as quite a different person.

Figure 1.1: Traveling in Bavaria—Summer 1960

My summer experience was powerful enough to convince me to take a year off between my Junior and Senior years in college to go back to Germany for a full academic year at the University of Bonn. This gave me time to learn the language well, to take more philosophy courses, to immerse myself deeper in the culture, to travel extensively, and to sort out what I really wanted to study and to do with my life. Even though I would complete my philosophy major after returning to Kalamazoo, I had to let my dad down about his vision of me as a protégé in the ministry. I just was not firm enough in my beliefs to be able to ascend the pulpit on Sunday mornings and offer parishioners what I thought they would want to hear. Aside from that, I did not feel comfortable speaking in public, and still don't after 45 years in higher education! Nonetheless, my family background and upbringing have profoundly influenced my values, and my attitude toward my profession and career. The fact that I like to work hard, that I have a strong service orientation, and that I wanted to help young people shape positive values and meaning for their lives, harks back to my childhood as the son of a Protestant clergyman. Though I am not an active church member, I nevertheless feel "called" to and driven by my career, which I take with the utmost seriousness, perhaps even with a missionary zeal. I owe that passion for service to my dad and my mainstream Protestant upbringing.

Though uncertain about my career direction, I had discovered for myself how extremely provincial and culture-bound Americans were in their knowledge and understanding of the world, and that this was a major handicap for us in the broader realm of international affairs and trade. I began to think, therefore, that I could perhaps make a difference through some kind of international work, perhaps even teaching. So, even though I graduated with a major in philosophy, I decided to build on my knowledge of German and Germany and was offered a fellowship for the Master of Arts in Teaching Program at Wesleyan University in Connecticut. Wesleyan was a great intellectual experience for me, where I took several courses in the German department as well as education courses through the MAT Program, and practice taught as a German teacher in Hartford Public High School. Totally unanticipated, the program also provided me with a full scholarship to spend another year as a student at the university in Würzburg, Germany, to be followed presumably by a career in high school German teaching.

The second year in Germany offered me greater depth of knowledge and better skills in German, more exposure to German literature, as well as new personal contacts and a growing life-long commitment to promoting and sponsoring dialogue between the United States and Germany. Though I had never really thought of myself as an academic, I learned about and landed a two-year lecturer position in German at Union College in Schenectady, New York, for 1965-67, and thus had a chance to teach the language to some bright young kids at a private, well-respected college and discover that I could impart to them not only knowledge of the language, but the importance of understanding the perspectives of people from nations other than our own. This was a powerful and formative time for me, which gave me the confidence to teach and to believe I could make a difference through my role as a potential professor of German. The two years went by very fast and made it obvious that further graduate work was in order, since I would need to have a Ph.D. if I really wanted to follow this path.

The next step was a quick three years at the Department of German at the University of Michigan in Ann Arbor. I knew that the degree would be in German literature, and that I would have to find and develop a specific area of interest within the prescribed canon of German literature from the Middle Ages to the present time, so I had a lot of reading to do and a lot of catch-up work in terms of literary theory and analysis. But that was fine with me, even though my first motivation was to teach the language and to figure out ways to get students over to Germany for experiences such as I myself had known. I enjoyed my research on the life and works of Franz Kafka and became thoroughly immersed in my dissertation, which explores the influence of an earlier German author, Heinrich von Kleist, on Kafka. I especially related to these two figures, since they gave such moving expression to the struggles of young persons feeling conflicted regarding their childhood roots. Yet, in the back of my mind, I felt literature, as interesting as it was and as much as it spoke to me personally, was nevertheless a detour around the work that I would consider to be most important. For me it was a frustration that I was being educated first and foremost to train other young people to be scholars of German literature. Pedagogy, program building, applied language learning, and outreach to a broader range of students were definitely not the first priorities in foreign language departments at that time.

Before explaining where my Ph.D. and my simmering doubts took me, I would like to jump back to the summer before entering the Wesleyan MAT Program (1963), which led me to much more than a closer connection to Germany. This was the time in my life when I would find a mate who would share my interest in things German and would be happy to go off to Germany for sabbaticals and other special occasions over the coming decades. Totally unaware of the long-term implications, Mrs. Ruth Wilson had contracted me that summer as a German tutor for her daughter, Carol, who was preparing to study in Vienna that fall for her junior year. Carol had been a neighbor in Ocean Park, Maine, where our families had adjacent cottages and where we both had worked in the summers. I had never really noticed Carol before this time, because she, being three and one-half years younger, was just the kid up the street. Now that I was 23 and she almost 20, however, the situation looked very different! As a result, our German lessons soon transcended language, and marked the beginning of a deep and lasting relationship that led to marriage two years later, just before I started my two-year teaching job at Union College, now 47 years ago. Carol, who had graduated from Colby Junior College and then completed her Junior year in Vienna with Colorado Women's College, was able to complete her undergraduate BA degree at Union as a faculty wife and then supported us both as a teacher for troubled kids while I earned my Ph.D. at the University of Michigan. We both enjoyed Ann Arbor and its big-ten culture immensely and still maintain friendships with fellow graduate students and their spouses and partners. We left Ann Arbor in 1970, with a Ph.D. and our six-week-old son, Peter, to start a new life and, for me, my forty-year career (or careers) at the University of Rhode Island.

While interviewing graduates of our program at URI, I noticed that many had found their mates through their international study experiences and that the common experiences of studying abroad and gaining a far more global perspective can easily help to solidify the bond of common concerns leading to a strong marriage. This has certainly been the case for Carol and me. We share German as a second language and a common interest in Germany and German-American relations. We have had numerous trips to Germany together, have spent two lengthy sabbaticals in Germany, and we share many friendships with Germans, of both a personal and professional nature. We often visit friends in Germany, and our German friends often join us in Rhode Island or at our cottage in Maine.

CHAPTER 2

My Unexpected Path to Engineering

As rewarding as my forty-year career as a German professor at URI has been, it has not always been that way. Having gone into the field with idealistic hopes of attracting lots of students to study German, to study abroad and to incorporate the language learning process and cross-cultural experiences into the core of their lives, I was soon disappointed by a low level of interest, indeed a growing hostility toward language learning and language study in the American academia of the 1970s and early 80s. Much of this was attributable to a new separatism and isolationism in the wake of the Vietnam War and the ensuing protest mood and "liberation" of the curriculum on American campuses. In any case, the number of German learners in American colleges and universities was shrinking and fewer and fewer students were interested in German for the sake of studying literature in depth, as I had done. My own situation was made even worse by some of the myopic thinking of colleagues who were entrenched in their own rigid pedagogical modes and did not want to encourage sending students abroad, since it would appear to mean even fewer students in their classrooms. I put myself in jeopardy by suggesting some changes to the curriculum, different textbooks, different testing measures, and so on. Despite my good intentions, the head of the German program became so upset with me and my ideas that she did everything in her power to deny me tenure, even though I had established a good teaching record and had begun publishing my research on Franz Kafka. In these early years at URI, I had discovered rapidly that something fundamental needed to change in the profession if language learning were to play a meaningful role for American undergrads, and that, for my own sake and personal well-being, I needed to either find the freedom to create change or leave the profession for other ventures which might be more fulfilling. There were times in my early career at URI when I was disillusioned, very discouraged, and just plain depressed.

At URI, I was confirming what I had suspected in graduate school, namely, that the traditional literary mission of a language department in American higher education had effectively limited its reach to a very small number of students. Literature studies, with its well-defined canon of prescribed authors and their works, had become the primary focus for advanced courses in U.S. language departments with the adoption of the Humboldt model of university education in the nineteenth century. In other words, American German departments had grown up and matured trying to copy the German departments in Heidelberg, Munich, Berlin, and other classic German universities. In my opinion, as valuable as literary studies might be, there was no logical reason for defining literature as the primary and sole end of the language learning process for American students. Being able to study German literature was logically one end product of mastery of the language, but surely not the only one. Would not some students want to learn German simply to travel or to enhance their careers as scientists, engineers, art historians, or musicians? Would it not make sense to expand the concept of German culture in the classroom to include topics such as science and technology, upon which the German economy is built? Would it not make sense to develop German classes with such topics? Though this might sound like a simple and common sense conclusion today, the literature tradition was rooted so deeply in language circles at that time that its wisdom was not even open for discussion. To doubt or challenge the literary curriculum in a U.S. language department was tantamount to heresy. I thus continued to teach the eighteenth century courses and the works of Goethe and Schiller to a handful of majors at URI, while my colleagues taught the nineteenth and twentieth centuries, hoping that some offspring would go on to graduate school to carry forth the model they themselves represented. This was all happening at a time when university budgets were constricting and deans and provosts were beginning to point out that major programs like German with small numbers of students were not sustainable.

Throughout my earlier years at URI, I had quietly initiated a number of outreach programs to other disciplines, particularly in conjunction with the College of Business. Based on my conclusion that language learning should be related to all fields and not just the study of literature, I encouraged business students and also chemistry students to study German in preparation for internships with companies in Germany. I had established a relationship, for example, with the leadership of American Hoechst Corporation, the Rhode Island subsidiary of the German chemical giant, began teaching German to their research chemists, and was able to convince them to place a URI chemistry major as an intern at their German headquarters each summer. I found such experimental efforts quite stimulating for both my students and myself and was pleased to be able to arrange such programs abroad for a small number of students. The feedback from both students and companies was excellent and it became clear that international preparation for professional school and STEM discipline students was needed and highly desirable.

Though I had never considered engineering as a potential partner for a department of languages, logic would have suggested such, even at that time. Germany had always been famous for engineering, and technology was at the root of their economy. Why not encourage American engineers to learn German as a tool for better interaction with that side of German culture? But I did not know the URI engineering faculty, who were located in another corner of campus and they seldom found reason to interact with humanities faculty. In addition, engineering students were the one group on campus who were not required to take any languages classes at all. Looking back, it is quite clear that there was no obvious or simple way to unite languages and engineering at URI or any other campus for a common cause, and that it would take some kind of special circumstance or unusual course of events to get things off the ground.

One factor helping to lay the groundwork for my collaboration with engineering was my own proclivity for things mechanical. I had always loved machines and working with my hands and had been obsessed with cars in my teenage years. I used to think nothing of repairing a broken transmission, of replacing a clutch, or modifying an engine with dual exhaust systems or multiple carburetion. And I still have a passion for automobiles today. Furthermore, my older brother was an engineer teaching at Worcester Polytechnic Institute (WPI), and I had often thought about following in his footsteps. Given this inborn fascination with things mechanical, it was thus really not a far reach for me to want to draw a connection between German and engineering. On the contrary, when the opportunity arose, I was ready for it. And that opportunity did fall in my lap, even if unexpectedly.

Figure 2.1: Fine tuning my 1936 Ford.

Coincidentally at that time, I, together with my engineer brother, had built a house on speculation next door to my own home in Rhode Island—with the idea of possibly leaving URI and starting a new career as a contractor, such was the extent of my professional discouragement at that time. Given the weak housing market in that year, we had also decided to put the house up for rent for a year or so, with plans of selling once the market improved. As fate would have it, we quickly found a tenant in a new hire at URI, who happened to be the new dean of the College of Engineering! His name was Hermann Viets, and, as the two "n's" in Hermann revealed, he had been born in Germany and raised in a German-speaking family. He furthermore strongly agreed that engineers would need global preparation if American technology and economy were to remain strong and competitive in the coming years. Hermann and I had a meeting of the minds during a classic American backyard chat, and subsequently formed a committee of language and engineering faculty to lay the groundwork for a new international program for engineering undergraduates at URI.

Though we explored several options, the basic idea of the URI International Engineering Program (IEP) emerged quickly as the dominant model. It would be a five-year undergraduate program, enabling students to complete both the BS in any one of the engineering disciplines and the BA with a major in a language, initially German. In the fourth year of the program, IEP students would be sent abroad for a six-month professional engineering internship experience. There were doubts whether students would be willing to commit to an extra undergraduate year, meaning not only more time, but more tuition, and there were doubts whether we could convince faculty in either subject area of the value of this direction. We knew, after all, that engineering faculty would prefer more time for technical subjects and generally saw little need for language work. And we knew that language faculty tended to be fearful of programs that might possibly suggest their department is in the "service" of a professional school. Languages were, after all, proudly housed in the sphere of the humanities, where the ultimate mission is to inspire literary scholars.

Despite the doubters, Hermann Viets and I were very excited about this plan, and we both took it forward with a great deal of personal commitment and passion. The program was able to evolve because of the clear engagement of key personalities who were eager to act when given the opportunity, and it certainly made a difference that one of us was a dean. Though I was more than ready and willing to commit my time to this effort, it would have been extremely difficult, if not impossible, in the early years without the enthusiasm and support of the engineering dean. Hermann's attitude was extremely positive about this venture and he played the key role in selling this idea to the doubters among the engineers, who thought the whole world spoke English and saw no need to "dilute" technology with language and culture study. I, in turn, had my work cut out in convincing colleagues in languages that it was not heresy to develop special language courses for engineers and to move the curriculum in German more toward a comprehensive German studies program versus a pure literature major.

It is important to note that the evolution of the IEP coincided very closely with the fall of the Berlin Wall and end of the Cold War, the opening of China and India to the free market system, the development and spread of the Internet, communication technologies like the fax, and the efficiencies of rapid travel. Hermann and I were well aware at the time that things were changing and that business, industry, and technology were becoming increasingly global and would demand that its future players be able to work internationally and cross-culturally. Our instincts told us that our idea was important, and this was soon confirmed by the rapidly evolving historical context. We had an appropriate idea for the day, at a time when few others in academia were considering these geopolitical changes, the soon-to-be surging wave of globalization, and its implications for the university curriculum. In short, we were ahead of the curve.

CHAPTER 3

Building a Network of Support

A critical next step for us involved securing external funding. We knew from the beginning that a new idea such as this would not find immediate support from any regular campus revenue stream. Our colleagues might well tolerate our idea of founding such a program, but not if it took money from existing already tight budgets. In short, we needed funding to support new sections of German for the engineers, travel to Germany to line up internship commitments, funds to develop materials for recruiting students, and faculty release time to make it possible to get the work done. We also needed the prestige of extramural endorsement of our idea in order to help sell it at home. With a financial commitment from the federal government, it would be much easier to move forward with revolutionary ideas. It would not take money from anyone else's pocket, and we could point out to any doubting faculty that we had survived a tough national competition for support and were now morally and legally committed to our project goals, at least as an experiment.

Based on the experiences some colleagues had had at other institutions attempting to internationalize professional school curricula, we turned to FIPSE (Fund for the Improvement of Post Secondary Education), a risk-taking federal agency supportive of new and potentially replicable ideas for higher education at the U.S. Department of Education. FIPSE, with the help of its program officers Sandra Newkirk and Michael Nugent, saw the value of our plans and funded our start-up, actually extending support for our ever-evolving initiatives for a period of more than ten years. Both Sandra Newkirk and Michael Nugent[1] were German speakers and humanists who understood the value of language learning and its relevance to broader educational needs in the evolving era of globalization. They also understood the myopic thinking of many language faculty and hoped themselves to help launch a new and more inclusive culture for the language curriculum in higher education. The IEP was fortunate to be able to approach FIPSE at a time when its review staff was sympathetic to our cause, and open to a reshaping of the mission of a department of languages. At the time, FIPSE was only funding about 65-75 new grants each year for well over 2,000 submissions, making a FIPSE award very competitive and prestigious.

Being supported by FIPSE had many ramifications beyond the purely financial. As a FIPSE grantee, I was invited to attend and participate in their annual conference each year of our funding, which was designed to support faculty with promising ideas, in part by making us aware of other reform efforts at universities around the country, and by linking us with like-minded colleagues at other institutions. Initially by way of FIPSE I laid the groundwork for my own network of higher

---

[1]Michael Nugent provides an excellent description of his own professional evolution as an "applied linguist" and advocate of languages across the curriculum in: Downey, Gary and Beddoes, Kacey, editors, *What is Global Engineering For?: The Making of International Educators*, Morgan & Claypool, 2010.

education reformers which would grow annually and is still growing today, even after my retirement. This would ultimately enable me, again with FIPSE support, to build a more formal network of faculty committed to the internationalization of engineering education and to found the Annual Colloquium on International Engineering Education (see Chapter 11). FIPSE conferences were invigorating, since every attendee represented his or her own idea and was committed to making it stand the test of success as a sustainable and replicable model for the improvement of higher education across the country.

As a note to those seeking external funding, I would like to stress the importance of persistence and determination. We were not funded the first year we submitted, even though we had made it to the last group of finalists. We did, however, go back to the Department of Education the next year, when we were successful. I recall my discouragement at the time due to the highly competitive nature of the FIPSE funding program. But, when asked by a senior colleague in our dean's office, Professor Robert Gutchen, if I was planning to resubmit the second year and answering negatively, he literally took me by the arm and sat me down at my Macintosh and insisted that I start writing. Had he not done that, as I myself have done numerous times since then for younger colleagues, the IEP, which has proven itself to be very fundable, may never have become a reality. Robert Gutchen helped me to understand that the grant writing culture, which was very much a part of daily life for colleagues in the sciences and engineering, could also be made available to humanists. Learning grantsmanship at this point in my career was invaluable and a key to the success of the IEP. This initial grant led to over a decade of ongoing support from FIPSE, as well as grants from the National Science Foundation, the Endowment for the Humanities, the Department of Defense, the German and Chinese governments, and several foundations.

My secret to successful grant writing, learned at that time, was to develop a concept which could be explained clearly and succinctly, even in an "elevator conversation." *The IEP is a five-year program preparing engineers to work globally. It includes a year of work and study abroad and leads to a BS in engineering and a BA in German.* Of course, a proposal must be far longer and must include far greater detail, but I quickly learned that success meant grabbing the attention of bleary-eyed proposal readers with something that made sense and could be rationally and easily explained. Ever since this first grant, I have always told interested potential supporters, donors, and funders that the IEP is not rocket science, but rather a straight-forward reconfiguration of existing curricular components as required by a new geopolitical and economic era, namely, that of globalization. In short, it was a curriculum designed for its time that would align the university with the needs of the private sector and open doors for our graduates. Given the simplicity of the concept and the power of its potential impact, it became, if not an easy sell, at least a program that made sense to just about all persons concerned with the future of higher education and its relationship to national economic competitiveness and security.

Our first grant from FIPSE gave us the prestige we needed on campus and the resources to take the necessary initial steps. We thus launched beginning sections of German for engineers in order to keep this special group of students together, build an *esprit de corps*, and integrate technical vocabulary and culture into the language learning process. We developed promotional materials and started the critical student recruitment program. And we began the corporate outreach needed to ensure that we would have internships for IEP students in their fourth year. Fortunately our initial recruitment efforts confirmed and reinforced our instincts. Over forty enthusiastic engineering students signed up for beginning German courses in the fall of 1987 to get themselves on track for the five-year BA/BS program, and especially the potential six-month internships abroad with global companies. Hermann Viets and I had convinced ourselves that we might have 12-15 students in the first class, not forty. Clearly, we were ecstatic!

Despite the clarity and relative simplicity of the IEP idea, we also knew from the beginning that the program would require more than a critical student mass and the commitment of the faculty from two rather disparate colleges. If we were to place students in professional internships abroad, which was our promise, and if we were to find ongoing financial help for the project, we would have to rapidly develop a partnership with the private sector, convincing them that investment in our students and the program would be to their benefit as well. Hermann Viets brought his considerable skills in collaborating with industry to the table, and I added to that the connections I had made with German-American companies in the area, and the German Consulate in Boston. Our common outreach to industry, which meant contacting both German subsidiaries in our area and local Rhode Island firms doing business in Germany, confirmed our belief that the need for globally prepared engineers in the workplace was strong and growing stronger. He and I visited many companies and business leaders together, in both the New England area and in Germany, and these early trips gave me the confidence to take our message to company boardrooms myself and to ask for support and help by convincing them that our undertaking was of clear mutual benefit. This took a bit of time for me as a humanist, who was not in tune with the fact that my field really had a market value and that we were on the verge of building something of lasting value, with very broad consequences. The key was to learn that I was not asking for help for myself, but rather for the education of students who would serve these companies well in the coming decades.

A key action in those early years was the formation of an IEP Advisory Board, comprised of globally involved business leaders from the U.S. and abroad, interested citizens and government leaders, all of whom would become committed to and passionate about the program. We were very fortunate at the time to be able to convince an influential Rhode Islander to join our effort as Advisory Board Chair. Heidi Kirk Duffy is the widow of one of Rhode Island's most successful industrialists, the late Chester Kirk, who was a URI engineering alumnus. She is a native German, long in the U.S., but nevertheless well rooted in Germany and well connected to several key players in German industry. Readily identifying with our cause, Heidi was able to help us establish contacts with German companies for internship placements as well as for long-term relationships. At the time of this writing, she remains as our only chairperson. She has helped us immensely, both financially and otherwise, and she continues to be a stalwart supporter and advocate, setting the pace for all of our board members. Above and beyond that, we have become good personal friends.

Figure 3.1: Advisory board members Heidi Kirk Duffy and Uwe Berner.

The program was successful in attracting several persons of influence to the board with a strong commitment to our goals and an eagerness to provide advice and support. The board meets minimally once per year, which provides, among other things, incentive and discipline for the program to prepare an exhaustive annual report of its activities and status. The board serves to provide feedback, expertise in strategic planning, financial support, and occasionally the members also lobby on the program's behalf. Members also help in the networking process as we need to reach out to new industry contacts or to government agencies or foundations on an ongoing basis. The board meets annually, often in the IEP House on the URI campus, but every two-three years we organize the annual meetings in conjunction with our university or corporate partners abroad, and have met so far four times in Germany, once in Spain, and once in France.

Building relationships with IEP board members and globally engaged companies both in the U.S. and abroad has been an extremely rewarding experience and has led to long-lasting professional and personal relationships which I still enjoy in retirement. Hermann Viets and I learned early in the process that cross-cultural communication, both in terms of language and culture, was a very real issue for companies, whether American companies engaged abroad or subsidiaries of foreign owned companies in the U.S. Thus, the leadership of such organizations and persons of influence in the broad area of economic development and trade immediately saw the value of educating American engineers to speak languages other than English, and they were prepared to help support the IEP concept. Over the years, therefore, we have been very successful in building relationships enabling IEP students to intern both in the U.S. and abroad, first in Germany, then also in Spain, France, China, and Latin America. And many of our graduates have gone on to full-time employment with our partner companies.

To sustain relationships such as this, it has been necessary for me, and now my successor, to travel minimally once a year to visit our company contacts, and to nurture the relationships with partners who have now received and mentored over 450 IEP interns abroad, and have then employed many of our 400 graduates. This is sometimes complicated by the fact that our personal contacts move on to other positions, or retire. And occasionally partner companies are bought out by other companies or even go out of business. The flow of e-mails is incessant to maintain such contacts, to rebuild changing contacts, and to develop relationships with other firms as well.[2] On the whole, visits abroad, even though they are intense with multiple appointments each day, are a pleasure. Often we are invited to dinner or lunch and sometimes even to weekend hiking in the mountains. Several of our contacts have been with us from the beginning, and, as mentioned above, have become our good friends.

Hermann Viets was the key inspiration for me in getting much of this started, even though he unfortunately did not stay long at URI, as he soon moved to the presidency of the Milwaukee School of Engineering. Hermann had applied to become the president of URI in 1991 and actually became one of two finalists. But the final decision for the position by the Rhode Island Board of Governors for Higher Education left him just one vote short, whereupon he felt it was time to move on. Hermann certainly left his imprint on the URI College of Engineering as a whole and did not leave without convincing the key players and decision makers in his realm of the value of the IEP. Hermann also carried the IEP experience to his new position in Milwaukee where he has developed exchange programs with partner universities in several parts of the globe.

At the time of Hermann Viets's departure from URI, I assumed the directorship of the program on my own and retained that role until my retirement in 2010, with its continually evolving list of responsibilities. I taught many of the courses, regularly made annual or semi-annual visits to German business and industry to build relationships and create internship placements for IEP students, and I nurtured the program as it built university exchange opportunities, and added French, Spanish, and Chinese. Grant writing and networking for fundraising purposes also became a key piece of my annual cycle of activities which would mean bringing in several million dollars of support over the

---

[2]But how thankful we are for e-mail. In the early stages of the program, communication depended on telephone and normal postage contact, gradually being upgraded by the fax. We sometimes forget that we still had typewriters in our offices into the 1980s!

Figure 3.2: Advisory board touring on the Seine as guests of France's Total.

years on behalf of the IEP and its students. Then, as we will see below, I also took the lead as we developed a two-building complex for our living and learning community known as the Heidi Kirk Duffy Center for International Engineering Education. Today we speak of our numerous business contacts as partners, many of whom have become avid supporters of the IEP and have reaped the program benefits by hiring our students.

It was very rewarding to those of us committed to the program in its earlier years, to be recognized for the value of the IEP from authoritative outside voices. Some of the engineering faculty had been concerned, for example, that ABET, the Accreditation Board for Engineering and Technology, might not approve of the steps taken to form the IEP. But when ABET came to URI in 1990 for its regular review of URI engineering programs, the team was very impressed with the IEP initiative and immediately nominated the program for ABET's national Award for Educational Innovation, which was awarded to us at their annual conference in October 1992. On the language and culture side, the program caught the attention of the German Consul General at the Consulate in Boston, who was impressed with the value of the program for advancing interest in the German language, and promoting good U.S.-German relationships in the areas of business and technology. For that, they awarded me with Germany's Federal Cross of Honor, its highest civilian award.

The program had also caught the attention of the leadership at the Fund for Improvement for Post Secondary Education, which, in turn, funded us for a second three-year cycle upon the expiration of our first grant. It was exhilarating for all of us involved in the IEP to be endorsed by authorities outside of Rhode Island. It was rewarding by itself at this time to successfully secure internships in Germany for our first students going abroad, and to receive positive feedback from them about their experiences. To also be encouraged by those involved with national higher education standards and reform seemed like icing on the cake.

CHAPTER 4

Sidetracked by a Stint in the Dean's Office

My life at URI was complicated in these early years of the IEP when the Dean of the College of Arts and Sciences invited me to join his team as Associate Dean for Student and Curricular Affairs. In this position I would be responsible for curricular and teaching matters such as the management of the general education program and the introduction of new courses for the college's 26 departments. I also dealt directly with students encountering problems that could not be handled within the departments themselves. My four years in this position, followed by one year as acting dean, provided a valuable learning experience, even if it did somewhat slow down the evolution of the IEP. As associate and then acting dean I gained an appreciation of the larger operation and issues of the university, which would serve me and the IEP well throughout the rest of my career at URI.

Joining the dean's office meant not only that I would be balancing several jobs at once as language teacher, associate dean, grant manager, and IEP director, but that I would also be asking myself which direction my career should take. Suddenly I could consider administration as a professional path, and I did indeed flirt with that idea for a while, especially when becoming acting dean during my fifth year in that office and I did decide to become a candidate for the regular deanship. As I look back, I am grateful today that I did not get the position, as painful as it was at the time not to receive an offer, since it no doubt would have ruled out any further development of the IEP. I felt confident that I had done a good job as acting dean during a difficult transition period for the college. The search process for the deanship, however, gave me a good taste of university politics. Based on a coalition of interest groups on the search committee (the women wanted a woman and the scientists wanted a scientist), I basically did not stand a chance. My fate was also sealed by the new president who made it clear that he wanted a new dean from the outside.

Though teaching, grant managing, program building, and administering became a tough balancing act during my time in the dean's office, my four years as associate dean and one year as acting dean nevertheless gave me invaluable experience in terms of making my way through the bureaucracy and politics of the university with the seemingly endless challenges associated with the IEP. In these years I thus became acquainted with and generally gained the respect of key people at all levels of the university, thereby building more and more allies for the IEP in the process. This even enabled me to follow my dean years as the chair of languages for six years, working hand-in-hand with administrators who had learned through my incessant proselytizing to recognize the value of applied language learning, i.e., teaching languages for all students and not just humanities majors, and recognizing language faculty research in non-literary areas. Becoming chair of languages made me realize that the IEP effort and the idea of applied language learning were paying off. Ironically, the person seen just a few short years ago as a heretic and traitor could now lead the department!

CHAPTER 5

Reshaping the Language Mission

The rapid growth of the IEP happily built a demand for German language classes that would not have been there otherwise and enabled me and the department to argue for new faculty at a time when the German program had been reduced to just two full-time people. This meant not only the opportunity to rebuild to faculty levels of years past, but also to reshape the program to the realities of a new kind of student. Being a part of the dean's team was very helpful at that time too, since it enabled me to convince the dean and other key administrators that literary publications should not be the sole tenure and promotion criteria for language faculty. When searching for new German professors, therefore, I was able to look for people who were not only sympathetic to the IEP, but who could bring specific teaching and research strengths to the concept of a German studies major versus a German literature major, and to specialized classes for engineers. We were very fortunate to find a colleague, for example, who had been a physics major as an undergrad before doing his Ph.D. in German. Two other colleagues had experience and expertise related to teaching German classes with business content and another colleague had sufficient science and math background to enable creation of intermediate language courses with a technical content. We became a team of five professors, all of whom value the IEP and are devoted to its students and willing to go the extra mile to support the students and to ensure the success of the program. Each year, for example, one or two faculty tour Germany with younger students in the program, visiting companies such as BMW and Volkswagen, and our partner university in Braunschweig, in order to give them a first experience in Germany and thereby encourage them to stay with the program and their German studies. If one looks at the resumes of our faculty, one finds strength in pedagogy, applied language teaching, and cross-cultural communication, with many excellent publications and conference presentations. Dr. Norbert Hedderich came to us from the Ph.D. program at Purdue, where he specialized in teaching pedagogy and had also developed an interest in teaching German for business. Dr. Walter von Reinhart, who earned his Ph.D. from Brown University in a more traditional literary area, had nevertheless worked earlier as a businessman, and brought a broad understanding of language applications in business and technology related fields. Dr. Doris Kirchner, who has just retired, had her Ph.D. from the University of Pennsylvania and brought to us both prior experience in teaching German to MBA-level students, but also the expertise to broaden our literature and culture offerings more in the direction of a German Studies Program. Dr. Damon Rarick has his degree in German literature from Brown University, but had first earned an undergraduate degree in Physics, enabling him to easily develop science-based German courses for our engineers. The program has also been fortunate to enlist periodic help from instructors from Germany supported by the German Academic Exchange Service. We are very fortunate to have a devoted teaching faculty, which is readily recognized by our students.

The early and swift successes of the IEP in German, including the FIPSE funding, the national award from ABET, and the Cross of Honor from Germany soon caught the attention of other language faculty at URI and those interested in other language and culture areas. The faculty in French began to discuss options with me in the late 80s and were strongly supported by the College of Engineering administration where Assistant Dean Richard Vandeputte was himself a speaker of French. I personally was concerned that a French option would draw from our same pool of existing students, thereby diluting our critical mass needed to be able to offer special language sections in German for engineers. Those fears were unfounded, however, as we soon learned that developing a French IEP would expand interest in the program and ultimately increase our overall numbers. The dean of the College of Arts and Sciences was also enthused by this idea and convinced that the next hire in French should be a faculty member devoted to the French IEP. Professor Lars Erickson was found at that time, a French literature Ph.D. who had studied chemistry as an undergrad and actually had worked in industry as a chemist before going to graduate school in French. Lars was attracted to URI, largely because of his enthusiasm for the IEP, and has been directing that arm of the program since then, and carrying out research in cross-cultural issues associated with the program. To support his efforts in the IEP, the dean of the College of Arts and Sciences provides him a course release each semester.

The next growth step was the Spanish IEP, which was launched with yet another grant from FIPSE and the hard work of Spanish Professor Robert Manteiga, who was followed by Professor Megan Echevarria. The initial impetus for the Spanish program came from William Silvia, one of our IEP advisory board members who felt we needed to be preparing American engineering students for work in Latin and South America, not just in Europe. We realized too at the time that Rhode Island had become a destination for many immigrants from Latin America and that a Spanish program could be a good tool for recruiting the best and the brightest of these students to engineering at URI. Students from these countries needed nurturing and guidance to understand that their native language skills were an asset rather than a handicap, as they are often told, and they could build on these skills in preparation for internships and study in both Spain and/or Latin America, and eventual global careers. Here again, Dean of Arts and Sciences Winifred Brownell supported the program with needed release time for the director of the Spanish IEP.

In developing these programs, IEP language faculty across the board expanded and redefined their research scholarship. Much of their writing and publishing is either pedagogical in nature and tied to the teaching of languages to engineers, or it is program- or curriculum-based, designed to explore and disseminate the concept of the IEP to language and engineering faculty across the country. The issues associated with training our students as cross-cultural communicators is also reflected in our research and publications. As mentioned above, this work is fully recognized by our administration for tenure and promotion. (See the Bibliography for an overview of publications on the IEP.)

It is fair to say that languages have flourished at URI since the advent of the IEP. The language faculty have experienced substantial growth and prestige in the past two decades in contrast to most language programs in higher education across the country, and have seen their department become much more central to the URI undergraduate curriculum. At the time of this writing, URI has 154 German majors, which makes it the second largest number of enrolled German majors in the country, approximately 145 French majors, a new rapidly growing program in Chinese with 55 majors, and a Spanish program with 162 majors. In total, the Department of Languages has over 600 majors at the time of this writing, which is a very impressive number for a small state university. The popularity of languages at URI and their good favor in the eyes of the administration may be traced in large part to the IEP and our overall notion that language learning must be rooted in many subject areas and tied to the phenomenon of internationalization, global awareness, and the need to keep the nation informed and competitive. American students will nod politely at the idea of learning languages to broaden one's horizons, but will not show up voluntarily in the classroom until they are convinced that it will impact their lives and strengthen their potential for career success. It should be noted that 25% of all engineering undergrads are majoring in a language, even though there is no language requirement in the engineering curriculum.

Most language faculty agree that the IEP has been good for the department and for the status of languages at URI. the status of languages at URI, and that the program has attracted much positive attention to the university as a whole. Yet, the relationship with language colleagues has not been without its tensions and a certain amount of ambivalence and even hostility among a vocal minority. There is still today a residue of philosophical difference for those who see languages as the heart of the humanities and therefore literature-based, with a fear of committing to anything applied, as it may suggest that we are in some way in the service of and thus subservient to other disciplines. There are also those who think that my colleagues and I have built a personal "empire" with the IEP and its two buildings across the street (see IEP housing discussion below) from the language building, and have far too much leeway and authority, indeed that we are soaking up far too many resources from the university. In recent years there have been, for example, a few colleagues critical of our efforts to build a Chinese program, fearing that any positions won for Chinese will mean positions lost from the traditional language programs. For me it is the ultimate irony that the only reservations at the university about building a Mandarin Chinese program have come from the language department. Now, as the university considers building a program in Arabic and possibly expanding Portuguese offerings, I am sure the same doubters will surface again and point their fingers of blame at the IEP.

CHAPTER 6

Struggling to Institutionalize

As the IEP grew, a major landmark in the program's history was the hiring of a full-time assistant director in support of me and the overall program. A national search brought Kathleen Maher to the program, a very capable and enthusiastic professional, who came to the IEP with excellent background in study abroad at the University of Iowa, with fluency in Spanish and a Masters in Latin American Studies. Kathleen brought the organizational skills deemed necessary for the creation of improved promotional materials, the design and implementation of a first-class quarterly newsletter, the expertise for helping to launch and manage an annual professional conference, the interpersonal skills to help run a residential facility for IEP students, for student advisement, and so forth. She also represented the first non-faculty professional for the program, and thus marked an important step forward organizationally. Unlike faculty who could serve for a period of time, rather than teach or serve in some other way, she was devoted 100% to the central administrative support of the program, and especially to school outreach. The IEP was fortunate to find the right person at the right time, and to be able to keep her on board for critical programmatic steps and development over a full decade. This all sounds smooth and logical, but getting the university to commit to this position and critical next step was not easy.

Kathleen had come to us first of all by means of a FIPSE dissemination grant, intended to help us continue to grow and spread the word to other would-be innovators. Having welcomed the grant as a step to help me with a program that was rapidly becoming unmanageable for one person (I was also chairing the entire language department at that time), the central administration agreed in advance that the university would support her position at the conclusion of the grant. Not surprisingly, however, and all praise for Kathleen aside, this position would not be funded by the university without a battle. Though I do not like to call on our advisory board for basic university issues, it took direct pressure from them on the central administration to secure the assistant director position when the two years of grant funding expired. Basically, the university was embarrassed into action by some of the IEP's external supporters, who were appalled at the lack of institutional support for what they knew to be such a stellar program. In defense of the administration, the institution was suffering with severe cutbacks in state support, and the deans and the provost no doubt believed I could do without this new supporting position. After all, I always had!

One might assume that the fight for the assistant director position was over once the provost agreed to fund it. But, in Kathleen's ten years with the IEP, the job had to be renewed annually and was not secured or made permanent until the point of my retirement. It seemed that the provost had funded the position, but had channeled the money through the College of Arts and Sciences, which, in turn, claimed that they had never received the funds. As a result, the position and Kathleen were thrown into turmoil annually, as the College of Arts and Sciences argued that they could not afford to cover the salary. When confronted, the provost's office always insisted that the money had been allotted. Despite Kathleen's superior work and the acknowledgment of all parties that the IEP was a raging success to the benefit of everyone, the university would not come through with a clear commitment. I still feel a knot in my stomach as I recall these annual battles, requiring us to waste so much time and emotion when there was much productive work to be done. It bothered me immensely that Kathleen could not be openly and unequivocally rewarded with job security. And it was personally embarrassing to have to tell our advisory board members that the university was once again waffling on its commitment.

A consistent challenge for the IEP has been the establishment of itself as a fully institution-alized program with adequate funding and personnel in place to maintain itself and take care of the needs of its students. It is worth noting that every one of our positions dedicated solely to the IEP was first established with outside funding, all from grants or donations that I myself secured. Kathleen was funded by a FIPSE dissemination grant for two years; two of our German faculty were sponsored in their first years by grants; our Housing Coordinator for the IEP living and learning community was funded by a grant when we added a second building, and is now supported fully by funds generated from the student housing payments; our first two Chinese language teaching faculty were supported by grants from China and the U.S. government, as was our Coordinator for the Chinese Language Flagship Program. I myself worked full-time right through the summers but received no internal salary support until the last three years before retirement. It is only in the past few years that the program has begun to receive regular support from the university, attributable in part to advisory board pressure, and in more recent years to the extraordinary enthusiasm for the IEP of Dean Raymond Wright of the College of Engineering. Prior to 2007, the IEP never had a university budget line which might have been there for routine office costs, travel to develop internship opportunities, or any of the expenses associated with this multifaceted program.

In one sense it is unfair to say that the university does not support the IEP. After all, my salary was always covered as a regular professorship in the College of Arts and Sciences. In the earlier years of the program, I was always able to "buy out" my time for the IEP with grant funds, but eventually the College of Arts and Sciences agreed, although not unequivocally, that my time was best spent on the IEP, and that supporting me to do this was a good investment on their part. Dean Winifred Brownell also directly supported the IEP by granting release time assignments in support of the French and Spanish IEPs. She has always been a "cheerleader" for the IEP and openly proud of the achievements of its language faculty. The failure to support the program adequately is largely due to its interdisciplinary nature and the fact that it is not organizationally rooted in either the College of Engineering or the College of Arts and Sciences, though its faculty are. I, for example, was never formally the IEP director on the university books, even though I acted as such, but rather a Professor of German. Though the program benefitted students and departments in both colleges, it was organizationally not really housed in either one and there was no real mechanism for pulling it out of its never-never land. This delay in organizational structure was complicated by the fact that internationalization was for many years not high on the agenda of the senior leadership of the University of Rhode Island, even though the IEP has always been praised by administrators on all sides. It also suffered from disagreements among the two deans as to who benefitted more and who should pay the bills. It has become clear that the program thrives on its interdisciplinary character, and yet, at the same time, this has given rise to administrative resentments, as one college would argue it bore more of the financial burden than the other. Even though interdisciplinarity is "in" today in academia, university organizational structures are not readily or sufficiently flexible to support its development when programs truly do cross existing disciplinary borders. In academia, we all lose from this inflexibility.

I recall my optimism about university support when a new president came to campus in 1991 who gave voice to strengthened undergraduate education and a "new culture for learning" in which students and faculty would be encouraged to cross the disciplines and learn experientially in new modes and configurations. When he solicited faculty for good ideas to help fulfill his vision, I went to him armed with publications and awards, and asked for his support for the IEP. But that bubble burst rapidly when he turned me down, citing the IEP as a complex, expensive, and basically impractical model. Having been involved at his last position in building a campus in Japan for the Minnesota state college system, which had been very costly and only a marginal success, he seemed to be soured in terms of doing any serious international program building. He apologized for not being able to support me, and explained that his notion of an international program at that time would have been, so he suggested, a cross-disciplinary program in peace studies.

Despite the undesired outcome of my meeting with the president, it did have one truly positive impact: It made me mad and caused me to dig in my heels and decide to push forward regardless, determined that we could find support through grants and private sector donations, with or without the university's help. I knew we had a good thing and was not going to let it slip away! As I told friends and colleagues, we needed to privatize. I even kidded with friends that we should perhaps put ourselves up for auction.

Indeed, we were fortunate to maintain ongoing support from FIPSE (U.S. Department of Education) for a total of eleven years, and when I took my plight to the IEP Advisory Board, they agreed that the IEP should go its own way to the extent possible, and they began at that time to open their pockets, providing the funds we needed to support students and further develop the program. The most notable commitment at that time came from Board Chair Heidi Kirk Duffy, who not only provided an annual gift for our day-to-day needs, but pledged a $1,000,000 life insurance policy to eventually support an endowed directorship.

The commitment of the board, of course, gradually caught the president's eye, who began to take a closer look at what we were doing, ultimately to realize that this program was not only good for engineering and languages, but also for the university as a whole. It was thus not long before the IEP began to surface in university literature and in presidential speeches illustrating the progressive programs of the University of Rhode Island. Despite the president's eventual recognition and verbal support, however, the IEP did not enjoy a university budget line until 2007, when it was finally pressured and embarrassed once again by the advisory board to provide an annual modest amount.

Financial support for the IEP has been an ongoing process and has come from many sources. Aside from FIPSE there have been grants from the Department of Defense, and the National Science Foundation. Other generous grants have come from the German and Chinese governments. There has been support from private foundations, from individuals, and from the companies and corporations with which we cooperate. More recently the alumni have also begun to give back. I am proud of the fact that I personally raised far more money over my forty years at URI than the University of Rhode Island paid me in salary and benefits!

Upon my retirement in 2010, a fund raising effort was established to create an endowment to fund the IEP directorship in my name. It seems that all parties agree that an independently funded directorship would provide the program with independence from the budgetary fluctuations of the university and would shield the director from the annual debate over who pays what to support the program. To date, the program has gifts and pledges of approximately $1,300,000 toward a goal of three million dollars. And the university has committed to make this fund one of its top advancement priorities.

CHAPTER 7

Partnering with Universities Abroad

While the original IEP limited the student time abroad to a six-month professional internship, IEP students today spend a full year abroad, first as exchange students for one semester at a partner institution and then as professional interns for six months with partner companies. Given the number of IEP exchange partnerships with universities in Europe, Asia, and Latin America, it is perhaps surprising that university exchanges were not a part of our original plan. The addition of exchanges first came about when the wife of President Bernd Rebe of the Technische Universität Braunschweig in Germany came across an article about the IEP in the *Chronicle of Higher Education* in 1989, and brought it to her husband's attention. He, in turn, passed this information on to Dr. Peter Nübold, head of the Braunschweig Language Center, who contacted me and suggested that we may have some common interests. Peter Nübold had been overseeing Braunschweig language programs, and as we had begun to do at URI, had been devoted to teaching languages with different contact bases, e.g., teaching languages for engineers. He had also spent a sabbatical teaching in the U.S., felt comfortable with both cultures and was very devoted, as was I, to supporting the exchange of U.S. and German engineering students. With Peter's encouragement, therefore, President Rebe urged Hermann Viets and me to visit Braunschweig when on our next trip to Germany, which we did in 1990.

President Rebe was a gracious host who understood immediately the unusual nature of the IEP. He knew that very few engineering students from the U.S. were studying German and that almost no engineering faculty were sending their students abroad for other than very short-term visits. He urged us, therefore, to explore the possibility of exchanging our students. With this first contact, a joint venture was begun between these two institutions, which has not only led to the exchange of well over 500 students, but to major research collaborations among faculty, exchanges of faculty at the two schools, as well as model dual-degree programs at both the masters and doctoral levels. The Braunschweig exchange has likewise served as a model for the creation of similar partnerships with schools in Spain, France, Mexico, and China.

It has always seemed ironic to me that I, the humanist German professor, would be a major player in bringing our engineering faculty together for transnational collaboration and the exchange of engineering students. Indeed, I remember wondering if I would be taken at all seriously by engineers on either side if I suggested their students be exchanged, or whether anyone would listen if I suggested we launch a Dual Degree Masters Program, leave alone Dual Degree Doctorate.

But these things have all come to pass, and the engineering faculty greatly appreciate the personal and professional connections that have been made with colleagues abroad. Indeed, this part of the journey has taken on dimensions I could scarcely imagine at the beginning of our relationship with Braunschweig.

The creation of the exchange with Braunschweig was aided initially by grant support from FIPSE and also a fair share of diplomacy. When the idea of an exchange with Braunschweig was proposed, by me on the URI side and by President Rebe and Dr. Peter Nübold on the German side, there were immediate questions from faculty doubters. Neither knew the other and, in most cases, had never heard of the other. "Why Rhode Island? Why aren't we talking with MIT or Cal Tech?" "I have never heard of Braunschweig. Why aren't we talking with Aachen, Munich, or Berlin?" Thanks to FIPSE, we had the opportunity to fund travel and to dispatch exchange visits of faculty delegations from and to both institutions to help resolve these doubts. The faculty could meet each other, gain an understanding of the way engineering was taught at the potential partner school, and obtain an overview of research interests and research facilities. With these visits and resulting personal contacts, the doubts evaporated, each felt comfortable with the other, and the engineering faculty at both schools gave their blessings to a student exchange.

The negotiations for details of the exchange were the second challenge, but were very much facilitated by the cooperative and creative spirit of Braunschweig's Peter Nübold, who was appointed by President Rebe to be Braunschweig's main contact person with Rhode Island and by Professor Jörg Schwedes, who was in charge of international exchange programs for the very large School of Mechanical Engineering at Braunschweig. Schwedes and Nübold set the tone of our meetings by stressing that such agreements must be based on mutual trust and respect, both personal and professional, and that no exchange such as this would work without flexibility in course and credit recognition. Peter Nübold and I drafted the exchange agreement based on the idea of a straight-forward one-to-one exchange, with students fulfilling their financial commitments at the home institution. Since this meant somewhat of a financial imbalance due to the fact that Braunschweig did not have tuition, with their students' costs covered completely by the State of Lower Saxony, he suggested that Braunschweig "throw in" a four-week intensive language course for the URI students prior to the beginning of the semester, with expenses paid by the German exchange students. This idea was accepted by all and was a key part of the negotiations for the first agreement between Braunschweig and URI, which was signed in 1995.

Peter Nübold and I became the point persons for the exchange which was launched immediately after the signing for the 1995-1996 academic year, and were responsible for all details, serving also as the primary hosts or contact persons for the students coming to our respective schools. In the process of taking care of these matters, Peter Nübold and I quickly gained an enormous respect for each other and quickly found that we could fully rely on one another. It helped that we were both language faculty and thus had much to share on the professional level. We also had each spent extensive time at universities in the U.S. and Germany and had a good grasp of the differences, both linguistic and structural, and could anticipate potential trouble spots. It helped immensely that we were both long-term faculty at our respective institutions and thus knew whom to contact or where to turn when there were problems. It is not surprising that Peter and I became close personal friends over the years as we nurtured the relationship between the two schools and took care of our respective students' needs. Eventually the program impacted not only engineering students, but also faculty and students in business, the biological sciences, pharmacy, and chemistry. At the time of this writing, the two universities have exchanged well over 500 students who pursued dual bachelor, master, doctoral degrees or an MS/MBA, not including any of the many short-term visits or programs.

Figure 7.1: Peter Nübold and I at the 2012 IEP reunion in Braunschweig.

Phase two of our agreement with Braunschweig involved the creation of a dual-degree program in 1997, through which TU Braunschweig students could fulfill all requirements for the URI Masters of Science in engineering and the *Diplom* in engineering at the Technische Universität Braunschweig and receive both degrees. The idea of such a program between a U.S. and a European university was basically unheard of at that time, but made enormous sense, since it provided graduates with clear credentialing at the graduate level in both Europe and North America, and included research in the form of a masters thesis that was accepted and approved by faculty on both sides. Graduates of such a program would be well grounded in engineering as practiced in both the U.S. and Germany.

One can imagine the barriers thrown into the path of such a revolutionary idea by faculty on both sides as well as administrators such as graduate school deans and registrars. The idea of the dual degree had begun with one of our Braunschweig exchange students, Mirko Kugler, who discovered he had the background to do grad-level courses in civil engineering at URI and thought maybe he could stay an extra semester or two and complete a URI Masters degree. After examining his record at Braunschweig and talking with some engineering colleagues, I decided to advance the idea of a two-year dual MS/*Diplom* program, with half the work done at URI and half the work done at the German institution. Students would complete the thesis at the host institution, to be approved by faculty on both sides, and otherwise satisfy all requirements of both schools and then receive degrees from URI and TU-BS. The idea was exciting, but brought out multiple naysayers who raised such issues as the following:

- "German students cannot apply for graduate work in Rhode Island, because they do not have the prerequisite bachelors degree!" It took long hours to explain that the German system had no bachelors at that time and that the first degree was at the MS level. Only after long arguments, especially with the dean of the URI graduate school, could we deal with individual cases and speak in terms of equivalencies.

- "ABET certainly will not approve this!" This was a false issue since students were meeting all existing requirements. But tossing out ABET is a standard knee-jerk reaction among engineering naysayers, seeking reasons to be skeptical about any new idea. It needed to be recalled that ABET had discovered the IEP when making an accreditation visit at URI in 1992 and that recognized it nationally with its Education Innovation Award.

- "I don't want my grad assistant leaving here after one year! He/she is working for me." This is a real issue, basically only satisfied when faculty from both schools are truly collaborating with each other, enabling students to progress on the same projects while abroad. When that happens, everybody wins, but it is complicated to reach that point.

- "State agencies and governing boards will never accept the idea of granting dual degrees." This was an outcry from both sides, which was quieted through lengthy approval processes at both schools, and the recruitment of support from key faculty, deans, and other leaders. Serious objections from governing boards never emerged, since the students were all meeting existing requirements.

The Braunschweig students helped to clear the path in Germany by arguing successfully for recognition of URI MS courses toward the German *Diplom* and the URI thesis at the home institution, thereby enabling them to complete their final year of the *Diplom* program based on the year in Rhode Island. The next step involved Braunschweig accepting the first-year MS work of Rhode Islanders as the basis for their year of study in Braunschweig, in order for them to likewise secure degrees from both schools. The key compromise point for both schools was the acceptance of a common thesis.

Final acceptance of the whole dual degree plan came down to very tense negotiations in Braunschweig where a few obstinate faculty leaders were saying this simply could not be done. At this point, with several URI faculty in Braunschweig, President Bernd Rebe, a very colorful and charismatic leader, invited us all to lunch, which included a few bottles of very good Italian wine. We then walked back to his office, sat down around a conference table, where he looked his faculty in the eye, and said: "Now, gentlemen, how are we going to make this happen?" As a result, the dual degree program was signed in 1997, and many students and faculty have benefitted from this program, which has also inspired many similar partnerships between other German and U.S. institutions.

Figure 7.2: With Braunschweig President Bernd Rebe in 1997.

A major lesson for me in this process and in the evolution of the extension of the dual degree concept to the doctoral level was the importance of personalities, trust, and common sense. I don't believe we would have succeeded in developing the dual degrees without the support of President Bernd Rebe, and it definitely would not have happened without the commitment and perseverance of Peter Nübold and myself. Peter and I were close in age, had both been in our positions for many years, and were both generally well respected by colleagues across our respective institutions. Both of us pushed forward and refused to accept "no" from people who could not see the value of the idea. It depended very much on our persuasion and, at times, insistence, and our ability to find others who would help overcome the obstacles. Our philosophy was to find ways to foster a spirit of give and take, to make it work if it made sense and was valuable for our students and our institutions. Everyone needed to see that we would all benefit from taking this decisive step.

Figure 7.3: Enthusiastic alumnae at the Braunschweig reunion.

The dual degree program has been a success, widely recognized, and even imitated, and has created wonderful career opportunities for our students. Examples are Braunschweig grads who have gone to work for prestigious companies like McKinsey or Boston Consulting in Germany, or URI grads who have become senior automotive engineers at BMW or higher level engineers at Boeing.[1] The program has also laid the groundwork for major research collaborations, funded in part by grants from the National Science Foundation. More recently, the IEP has spawned a parallel program in business, the IBP, or International Business Program, and faculty in pharmacy have also begun exchanging students in support of joint research operations.

At this point, the exchange continues to flourish and will no doubt impact another 500 students. It does face new challenges, however, as the original model now must adapt to major curricular changes in Germany due to the conversion from the traditional *Diplom* to a bachelor/masters model. It is ironic that this has complicated the exchange, when its original intent was to simplify, but it has done so. German technical universities have developed a three-year bachelor, followed by a two-year masters. Though the outcomes are similar, they arrive at that point in quite a different way, rendering the old dual degree agreement useless. The two institutions are eager to address this issue, however,

---

[1]To read about the educational and career trajectories of such students, see: Grandin, John, *Going the Extra Mile: University of Rhode Island Engineers in the Global Workplace*, Rockland Press Rhode Island, 2011.

and an agreement of cooperation relative to the dual degree programs was signed by both presidents in 2012. Precisely how this is done will be the topic of future papers and conference presentations.

Though it sounds egotistical, the management of the dual degree programs has become more difficult following the retirements of Peter Nübold and myself, as well as some of the key faculty involved in the initial years of the collaboration. Such programs involve a keen understanding of the differences in the two systems and a concomitant ability to think in broad equivalencies and in terms of outcomes rather than, for example, insisting on precise course per course agreement. While maintaining high standards, it is nevertheless necessary to think creatively and flexibly. One needs leadership such as that of President Rebe who saw the value of the concept and asked not if, but rather how we could make it work.

CHAPTER 8

Going into the Hotel and Restaurant Business

As a result of the disintegration of campus residential fraternities, URI's centrally located Sigma Alpha Epsilon house, right across the street from the language department and close to the main entrance to campus, found itself vacant and without a purpose. The brothers had gotten into trouble with drugs and alcohol and had been indefinitely suspended from campus, leaving the alumni fathers of the organization discouraged after several attempts at revitalization. Even though badly deteriorated, the building had a solid shell and stood on an ideal central campus location, with easy access to classroom buildings. When given the opportunity to bid for its use, therefore, and bolstered by my own one-time building contractor experience, I leapt at the chance to create a campus-centered administrative and residential facility for the program, soon to be known affectionately as the IEP House. I was convinced and encouraged by several students that IEPers would love the idea of theme-based housing for the program and that they would benefit from mutual support while making their way through this rigorous curriculum. Though one might question the sanity of any faculty member wanting to create housing for forty undergraduate students in the building where he/she has his/her office, I was convinced that a home for the IEP and its students would create a special place for positive student/faculty interaction, and that it would help the program thrive. And I have never regretted this decision.

Strongly encouraged by the IEP Advisory Board and corporate partners who provided seed monies to explore the idea, and gradually endorsed by the somewhat skeptical university administration, I was able to put a financial package together to launch a $700,000 renovation, creating new program offices and housing for 40 IEP students. The fraternity would retain ownership of the building, assume a $700,000 mortgage for our renovations, and our monthly rental payments to the fraternity organization would pay down the loan over ten years. It was definitely a win-win situation. The fraternity would have a good tenant and substantial improvements to their building; the university would be able to eliminate an eyesore at the entrance to campus; the IEP would have a home for the program; and the students would have a great place to live together with fellow IEPers at a cost less than that of the URI dormitories. As good as it sounds, it was not a simple process, but after a tension-filled eight months, which required a great deal of personal involvement, costing me several nights of sleep and most of any summer vacation time I might have had, the IEP House was occupied by myself and 36 students in the fall of 1997.

Figure 8.1: IEP House residents.

To give one example of problems along the way, it became clear toward the end of the summer of 1997 that the house would not be ready for the beginning of the fall semester. We had been delayed by cumbersome state approvals, multiple inspections by various agencies, and the discovery of soil outside the building contaminated from years of fuel oil leaking from the building's underground tank. Furthermore, the construction of the front stairs and handicapped entrance, which were to be financed and built by the university, were way behind schedule. I, of course, had promised the rooms to the students as of September 1, and they had made payments to the IEP accordingly. What to do??? Well, it seemed that the fraternity next door to the new IEP House had also been basically abandoned and was scheduled for removal. So, with university permission and some quick short-term maintenance attention, we moved our new furniture into the shabby structure next door over Labor Day weekend. The students took this on as kind of a camping adventure and were amazingly tolerant of the situation, despite the bad condition of the temporary quarters. (That house would soon be razed.) Finally, in late October, just in time for the heating season, we were able to move our new furniture once again and establish life in the newly renovated IEP House, which exists since that time as a financially self-supporting, independent campus facility for the program and its students, including both URI IEP students and several exchange partners from Europe, Latin America, and

Asia. The IEP manages the house itself, and pays the bills by means of student rents and income from special summer intensive language programs. The students, who take great pride in the IEP House, govern themselves by means of an elected house council. I have always been proud of the fact that the house could run without a residential staff member. There have been a few alcohol-related incidents which have required some warnings and assistance from university experts, but basically the students are proud of being IEP students and understand that their status carries behavioral responsibilities. After all, the program is tough, requiring the students to use their time for study and not so much for partying!

Our original agreement with the SAE fraternity was for a ten-year period, with a clause assuring potential renewal beyond that time. At the ten year point in 2007, the success of the IEP House as a model living and learning community for the campus led to a university purchase of the building. Thus, the fraternity has left campus and the IEP remains indefinitely, with the program paying its monthly rent to the university, rather than to the fraternity. The university was pleased on two fronts. They finally succeeded in permanently eliminating a misbehaving fraternity from the campus's main entrance, and in holding on to paying tenants whom they could easily brag about. It is also worth mentioning that the success of the IEP House inspired further such houses on campus, a Women's Center, and a Rainbow (diversity) House, as well as several theme dormitories or dormitory wings.

Beginning in the second year of the IEP House history, I, together with the students, decided to equip and reopen the downstairs kitchen and dining room which had formerly served the fraternity brothers. I found myself in the summer of 1998, therefore, scouring the campus and other places for suitable kitchen equipment to be able to serve three meals a day for 50-60 people. I recall finding, for example, a very nice steam table and a commercial-grade mixer and some other equipment in the abandoned kitchen of the house next door which was about to be razed. And I remember finding a commercial gas cook stove in yet another abandoned fraternity. All they needed was a moving crew, which included me, a good cleaning, and a professional tune-up to be put back into service. I also conducted a search for a chef and hired Mark Schoenweiss, a highly talented cook with his own pedagogical spirit, who has endeared himself to the IEP students and the IEP community over the years, and spoiled us with his outstanding cooking. Mark is a very thoughtful, self-educated man who recognized what I was trying to accomplish through the IEP meal program, and he jumped in with enthusiasm, helping to educate me about equipment needs and how to serve three meals a day to all the residents and other IEPers who wanted to be part of the meal program. Mark still cooks for all residents in both of our houses today, as well as for hungry faculty and other non-resident IEPers. All enjoy the food at a bargain price as well as the camaraderie of being able to share the lunch or dinner table with others who speak the same language and are committed to similar personal and professional goals. Mark takes it upon himself to make sure IEP students develop a global culinary sophistication appropriate to the careers for which they are preparing. It is not unusual to find "duck" a la…. or "quiche" as part of a summer-time dinner or lunch menu which is, for Germans, a very special meal usually served over the Christmas holidays, and which is also a staple of the Chinese

Figure 8.2: They wanted to eat, too!

cuisine. Mark cooks with an eye on the culinary preferences of young American students, but also their exchange students from abroad. One can argue that this is the first immersion not just in the language of the country the students will eventually spend a year abroad in, but also in its "culinary" culture.

I would like to honor the creative role that the URI Fraternity Mangers Association and its director, Scott Tsagarakis, played in the creation of the IEP living and learning community. Scott was closest to the problems in the fraternity system and was alert to the fact that the IEP might be able to fill a role in the changing campus culture and related need to compensate for the failing fraternity system. Scott was the one who first brought the suggestion of an IEP House to me, and he has played key roles along the way in finding the right contractor, understanding the relationship with the university for a privately owned building on campus property, in eventually setting up and managing a kitchen, and dealing with many other aspects of "hotel and restaurant management."

I would also like to recognize the crucial role played in the development of the IEP living and learning community by J. Vernon Wyman, Assistant Vice President for Business Affairs at URI. Vern exercised enormous patience in dealing with a very unorthodox project for any university campus. He supported me from the beginning and helped to nurse me through the many steps with all of the outside agencies and officials. He became a "convert" and thus very committed to the project and understood how to keep it moving when the odds seemed very much against us.

Given the success of the IEP House as a home for the program and the continuing downward spiral of the American fraternity system, it is perhaps not surprising that the Chi Phi House next door to the IEP House also became available after their suspension for bad behavior. Though the challenges of creating the first house might make one leery of trying again, we did ultimately create a second home for the program. The idea of a second house sounded at first like an unrealistic dream, because it too would need major work and significant funding. But our board chair, Heidi Kirk Duffy, invited me to her home one afternoon where she and her husband informed me that she would donate her pleasure yacht to us, and the proceeds from its sale could be used to purchase that building if we could find the means to renovate it!

Figure 8.3: The former Chi Phi House - It needed a lot of work!

The IEP leadership thus soon made plans to purchase and update that building and incorporate it into a new two-building Center for International Engineering Education, which would be named for Heidi Kirk Duffy. Despite our initial euphoria, however, reality soon set in when we learned that the renovation would require a total "gutting," and that this, in turn, would bring a whole new set of requirements. Extensive renovation would demand special permitting and all of the features of a new building. Creating a new commercial grade kitchen, for example, would be a complicated and very expensive task, and the building would require an elevator to all four levels, also very expensive. Even though we owned the building outright, our renovation costs rapidly grew to $1,275,000 and beyond. This information cast clouds over our hope for the second building, but I was convinced that we should and could make this happen despite the enormity of the challenge.

Figure 8.4: The Texas Instruments House—dedicated September 2007.

The second house project could be the basis for a book in and of itself. But let me just summarize some of the key challenges:

1. The Chi Phi organization was willing to accept our purchase price and fully vacate, but only if the university would provide them with a building lot for a planned new house on Fraternity Circle, where most URI fraternities are now located and where their organization could be housed once the suspension was lifted. The university bought into that idea, but then discovered that the only available lot would need substantial work to deal with drainage issues. I have to give major credit to the university officials who agreed this transition was wise and who found the funds to address the lot requirements. Even though it all worked out, this issue set us back a year.

2. Assuming control and responsibility for a building on a state university campus is not a simple matter, and involves multiple levels of approval. Once the university authorities agreed, then

there were the state properties commission, the state building inspector, the state fire inspector, local town authorities, the state health department, and the RI Board of Higher Education, and, believe it or not, both the governor and the state legislature. All of this required detailed plans from an architect at a cost of almost $50,000, careful coordination with a contractor, and a clear explanation of how the IEP would use the building. And, of course, the costs for all of these plans had to be born up front by the IEP!

3. Not far into this project I found myself way in over my head in terms of the financial commitment. Our proposed operating budget would enable us to secure a loan and handle monthly mortgage expenses up to around $800,000, but the remaining costs, which had mushroomed to another $800,000, would need to be raised from donors. So, I quickly needed to reach out or give up.

After much thought, I came up with a plan to re-channel a recent significant gift for an IEP endowment into a fund to support the house renovation. I believed the donor would like this idea and we could offer him the chance to name the building after his company, Texas Instruments. My problem was that I knew the engineering dean at that time would not want me to request that of the donor, nor would the URI Foundation like to see an endowment used as expendable brick and mortar funds. The latter believed that endowments were untouchable and the dean thought he could get more money from this donor for his own projects. I was really on shaky ground with this idea, but felt it would save our project and that I should contact the donor with my idea anyway. I knew how strongly he believed in the IEP and how unlikely he would be to support something else at URI.

The result was as predicted. The donor loved the idea, and even added additional funds to the original gift, and the dean was angry, as was the president and the URI Foundation. It is not that the president did not like the idea of a Texas Instruments House and a Heidi Kirk Duffy Center for International Engineering Education at the entrance to campus; rather, he did not think that faculty should be approaching donors on issues such as buildings and naming rights, which I fully understand and accept. Well, all of this came to pass and the building is now renovated and complete and the infusion from Texas Instruments encouraged other major donations of over $500,000 for the project. Today everybody is happy, and the donor has given even more to the IEP. I would do the same again if necessary, despite a month or so of really bad sleep. Academic entrepreneurialism can be lonely, even if rewarding.

Fortunately there would be other major donors to this project who should be acknowledged: The Max Kade Foundation provided $200,000 to establish a German-language floor in the TI House, where German IEP students could live together and make quicker progress with their language skills; ZF Friedrichshafen AG, a major German supplier to the automotive industry, gave us $200,000, half of which was dedicated to equipping the new kitchen; the National Science Foundation provided funds for the video conferencing center, enabling smooth communications with partner universities abroad; the university took care of landscaping, sidewalks and parking lot repairs, as well as internet wiring; and the balance came from smaller donations from alumni and individual supporters.

A final hurdle in the path of actual construction was approval by the State Properties commission and the RI State Legislature. While this seemed at first to be very straight-forward, the commission did not make its deadlines and the legislature went out of session just before our demolition crew was to start gutting the house. Rather than wait another six months, an IEP board member spoke to the governor who decided that he could approve this while the legislature was absent. But the Democratic legislative leaders learned about this and screamed "bloody murder" about a Republican governor overstepping his bounds. Thus, we were caught in the middle of a political battle and read about our house and the IEP on the front page of the paper, fearing that the whole project would be torpedoed.

After all the *Sturm und Drang*, the Texas Instruments House opened its doors in 2007, and its dedication on September 28 coincided with the twentieth-anniversary celebration of the founding of the IEP. Thanks to its donors and support from the National Science Foundation, the building created more office and meeting space, a new kitchen and dining room, a video-conferencing center, and housing for 37 additional IEP students, as well as space for short-term visiting scholars. It boasts a Max Kade German Language Learning Community on the second floor, a Spanish language floor, a Sensata Technologies Living Room, and the most modern of telecommunications and networking capabilities, enhancing communication with partner universities throughout the world and supporting distance learning programs. In thanks for her years of support and her major gift enabling us to purchase the second building, the two-building complex is now known as the Heidi Kirk Duffy Center for International Engineering Education.

An additional important step upon the establishment of a second building was the creation of a full-time housing coordinator position to manage the two buildings. As mentioned above, the two-building complex is operated by the IEP as an IEP living and learning community, fully independent of the university housing and dining system. It houses almost 80 students, provides three meals a day to residents and others associated with the IEP, deals with most maintenance and upkeep issues, recruits its own residents, and offers its own appropriate program. The center has its own operating budget and income stream, which must be carefully monitored. Thus, in 2007, as the second house came on line, we created the housing coordinator position and hired Angela Graney who has done a wonderful job keeping track of the many sides of this part of the IEP, and has endeared herself to the residents. Par for the course, there was no university funding for such a position. Thus, Angela's salary and benefits are generated from housing income as part of the overall budget.

When I look back today at some of the steps I took to move the TI House project along, I shake my head in disbelief. Basically, I am not an aggressive person, and I often seek ways to avoid conflict. Yet, in this case, I saw what needed to happen for the greater good to come about, and I pushed forward, regardless of the potential turmoil. Throughout my history with the IEP, I often found myself in conflict with the status quo and in need of taking steps to overcome the limits of the system. I believe that people were basically willing to accept or tolerate my occasionally unconventional actions and my perseverance, because they could see in the long run that I was acting for the program and our students and not for my own personal aggrandizement. Of course,

Figure 8.5: A living and learning community.

my entrepreneurialism also played a role in granting me some freedom to act, since I was making major changes to the campus at very little cost to the university!

CHAPTER 9

Taking the Lead Nationally

In considering why the IEP "happened" at the University of Rhode Island and not at a larger or more urban institution, it is important to note the relevance of its place as a research and teaching university vis-à-vis other schools across the nation. Being a smaller state university with a strong commitment to undergraduate education enabled us to move forward with a student-oriented, programmatic idea such as the IEP when other universities would have seen such a labor-intensive program as a distraction from traditional faculty research. This is not to say that URI does not pursue or value classic research in engineering or the humanities. It is to say, however, that its smaller size and commitment to undergraduates allowed a freedom of action that would have been more vigorously discouraged at larger and more "prestigious" institutions.

Over the years, the IEP has gained considerable attention nationally and internationally as a model for internationalizing engineering education and is often consulted by others who hope to take similar steps. URI faculty are thus often on the podium at engineering, language, or international education meetings, and are often asked to contribute to publications. The program and its faculty have received national awards and recognition from the Accreditation Board for Engineering and Technology, the National Academy for Engineering, NAFSA, the Institute for International Education, the Association of Departments of Foreign Languages (MLA), the National Association of State Universities and of Land Grant Colleges, the German Academic Exchange Service, and the German government. It has also been featured in numerous publications, most recently in the *Chronicle of Higher Education* (May 18, 2012).

As a result of years of outreach and dissemination, the program gathered an extensive network of contacts with faculty at other U.S. institutions working toward similar goals. I had often consulted with these "fellow travelers" from Georgia Tech, MIT, RPI, WPI, the University of Cincinnati, and others about the possibility of meeting to exchange ideas, to learn from one another, and to share our collective wisdom with those who would also like to prepare their engineering students for experiences abroad. Thanks to encouragement and financial support from FIPSE once again, I was able to call a group of approximately 30 such persons together in the fall of 1998 for a two-day meeting in Rhode Island, which ultimately gave birth to the Annual Colloquium on International Engineering Education. By the second meeting a year later, we had gathered 50 persons and over 80 by the third year. Now in its fifteenth consecutive year, the colloquium has evolved into a regular and respected annual conference, attracting at least 150 engineering and language faculty, deans, provosts, international educators, and members of both the public and private sectors. The colloquium is a distinctive and unusual conference in that it brings together a truly interdisciplinary group each year, representing all constituencies invested in the internationalization of engineering education.

We take pains to point out the difference between this annual colloquium and other international colloquia for engineering which tend to be focused mainly on engineering education issues per se, and not the international preparation of engineers for global work. While the conference was held initially in Rhode Island each year, it now meets every other year at a partner university elsewhere in the country, where the engineering faculty have also sought to address this issue. The colloquium has thus met to date at Georgia Tech, Purdue University, Iowa State University, and Brigham Young University, and will meet in 2013 at the University of Kentucky.

The colloquium has had excellent success in attracting keynote speakers to address our concerns from their perspectives as educators, governmental or private sector leaders. The presidents of the University of Rhode Island and the University of Cincinnati, the engineering deans of MIT, Georgia Tech, Notre Dame, Brigham Young University, Iowa State University, the CEO of Siemens in North America, leaders from BMW, from Continental AG, from ZF Friedrichshafen AG, from FM Global, the German Consul General from Boston, the senior U.S. Senator from Rhode Island, the President of ABET, the Science & Technology Advisor of the U.S. Secretary of State, and many others have spoken at the annual colloquium, showing their support for the globalization of engineering education.

Figure 9.1: URI President David Dooley addressing the 2012 colloquium.

The organization of the annual colloquia has, of course, not been simple and not without growing pains. A meeting like this requires attention to detail and promotion throughout the year and can only be successful with a good program and an impressive list of speakers who are addressing issues of direct concern to faculty working on these matters at their home institutions. We have had to learn to secure good locations, which are conveniently reachable by air and rail transportation, and where faculty can attend at competitive prices. In Rhode Island we have been able to work with the Hotel Viking in scenic Newport, Rhode Island, which is very comfortable and provides excellent meeting facilities at a reasonable cost. Annually we fret, of course, since we are required to sign a contract with the hotel committing to a minimum number of rooms per night. To put it bluntly, if we did not offer a quality program and people did not attend, we would face a major financial crisis. But, the self-sustaining colloquium has been a success each year and we expect it to remain in place for years to come as the major American forum for the exchange of ideas on global engineering education.[1]

Another initiative putting the IEP in a national leadership role involves the creation and development of the *Online Journal for Global Engineering Education* (OJGEE), a publication to which more and more faculty are turning as a publication outlet for their own work and/or for information about the work of others. The idea of such a project had been on my mind for some time, and I turned to my colleague in German, Damon Rarick, for help, with hopes that he would take on this challenge. Damon had always been a superstar in computer related things and had been very helpful in taking the IEP web site to a more mature level. As I had anticipated, Damon learned not only how to create such a journal, but how to do it in a professional and first class manner. Visitors to OJGEE (http://digitalcommons.uri.edu/ojgee/) will find a well-organized series of papers, with a history of six volumes and two special editions. The journal has had some growing pains insofar as Damon was told by the administration that his tenure and promotion decision would not be strengthened by his work on the journal, i.e., that his efforts to create this journal would not be considered a valid contribution to his research portfolio. Despite this setback, which essentially crippled the work on the journal for a period of time, I feel safe in judging the project a success based, in part, on the fact that my own publications in OJGEE have been accessed and downloaded over 3,000 times.

[1]For information on the Annual Colloquium, see: http://www.uri.edu/iep/colloquia.

CHAPTER 10

Building the Chinese IEP

Thanks to the requests of the students themselves and the encouragement and wisdom of IEP advisory board members and other associated business leaders, the IEP took positive steps beginning in 2005 to create a Chinese component of the program. It was clear to all that business and engineering were moving to Asia and that no one could speak of preparing engineers globally without being able to teach Chinese and send some of their students to China for exchange experiences. With seed money from Sensata Technologies and its CEO Thomas Wroe, and a soon-to-follow endowed scholarship fund from the R.I. based Hasbro Corporation, both coming to us by way of URI alumni, the IEP leadership moved as swiftly as possible to establish Mandarin language teaching at URI and to create both study abroad and professional internship opportunities for IEP students in China. We moved forward with this even though no Chinese was being taught at URI at that time.

With the help of a native Chinese colleague in URI's Graduate School of Library Science and Information Studies, Dr. Yan Ma, I was able to make contact with the leadership for educational programs at the Chinese Consulate in New York and subsequently with the Ministry of Education in Beijing. Thanks to the fact that these officials were enthusiastic about the possibility of linking language learning with engineering education in the U.S., I was able to submit a proposal and receive funding from Beijing to support a full-time faculty member in Chinese at URI for the initial three years of that position (2006-2009). Shortly thereafter, I collaborated again with Professor Ma who spearheaded the drive to have URI designated as a site for one of Beijing's U.S. located Confucius Institutes, thus making URI and the IEP an important regional center for Chinese language and culture matters in Rhode Island and New England.

Again with Professor Ma's help, we were soon able to establish a relationship with one of China's very top engineering universities, Zhejiang University in the city of Hangzhou. We visited China together and laid the groundwork for both summer-intensive programs and long-term exchanges, both of which are now in full operation. As of 2011, the IEP is in its sixth cycle for summer intensive language training in Hangzhou. 2007–2008 marked the first full-year experience for engineering students studying for the fall semester at Zhejiang University and then completing six-month corporate internships following that semester. In 2011-2012, we had twelve students studying and interning in China, representing engineering, business, biology, journalism, and fashion merchandising.

The Chinese IEP took a significant step forward in 2008 when I successfully competed for $1,000,000 in funding from the Department of Defense's National Security Education Program

Figure 10.1: Meeting with officials of Zhenjiang University in Hangzhou.

(NSEP), enabling us to join the prestigious Chinese Language Flagship.[1] NSEP funding and the Flagship are intended to support attainment of the highest proficiency standards for American students studying the language and it has commissioned the IEP to create curricular and programmatic mechanisms to encourage significant numbers of American engineering students to meet these standards. The IEP envisions significant growth for its Chinese program over the coming years, as it works with local schools to create a seamless Chinese language learning pathway through the high school and college years, while strongly encouraging its coordination with the goals of the STEM subject areas. The Language Flagship seeks to educate fully bilingual professionals to work competently in the global marketplace.

The development of the Chinese program at URI, which was approved as a full major in 2011, happened with incredible speed, but not without pain and frustration, only some of which I will explain here. As I look back, I see that our good fortune in securing funds from China, from Sensata Technologies and the Hasbro Corporation, in winning the grant from the National Security Education Program, in building a partnership with Zhejiang University, and in securing a Confucius Institute at URI enabled us to move forward almost too rapidly for the culture of

[1]For further information, see: thelanguageflagship.org

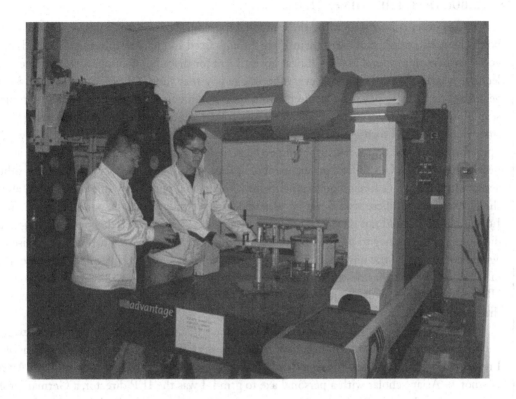

Figure 10.2: Curtis Richards—our first engineering intern in China.

an American university. Ironically, the resistance to a Chinese program came primarily from the Department of Languages, which, being made up of small individual language sections (e.g., the French Section, the Spanish Section), is designed to protect the home turf first. Rather than see the addition of a Chinese program as an enhancement of the departmental offerings and an expansion of the departmental footprint within the university curriculum, therefore, some of the more outspoken members of the department saw this as a threat to their own turf, which had not been adequately authorized. They felt that Grandin, with the assistance of external supporters, was pulling a power play and that he needed to be stopped. Fortunately, the broader university welcomed the addition of Chinese to its offerings, and the dean, the provost, and the president believed that this was the correct path to follow. Thus, even though I went through some humiliating meetings in the language department, and took a lot of abuse from some narrow-minded colleagues, I believe today that almost everyone has by now accepted the fact that China is important and that it is good to be teaching Chinese at URI. It was certainly comforting to us all in the IEP when the program found very strong endorsement from Dr. David Dooley who arrived as the eleventh president of URI in 2009.

One other episode regarding the Chinese program is well worth telling, since it illustrates so boldly how institutions can harm themselves through fixed and narrow bureaucratic mindsets. I was ecstatic when my proposal to the Chinese government for the funding of our first full-time language lecturer was approved. Just about everyone was happy that the Chinese would give us $40,000 a year for three years to pay for a lecturer whom they would help us to find through an international search process. But, after hiring the lecturer, when beginning the process of transferring the funds from Beijing to URI for the first year's salary, the vice provost's office declared that this was, in his opinion, a grant, and that Grandin should go back to the Chinese to get overhead monies at the rate negotiated with the U.S. federal government of at least 46%. Given the fact that URI had otherwise accepted support from foreign governments for instructional purposes without overhead, and given the fact that the Chinese would not be prepared to even understand the concept, I explained that I could not and would not do this. The dean's office then joined the discussion, declaring that they did not have the money to pay for the lecturer's health insurance, which had been promised in the agreement signed by the university. "Patience, John! You got the attention of the Chinese Ministry of Education, which was destined to provide much more support down the road, and you secured URI's first full-time faculty member in Mandarin Chinese for the cost of health insurance, and now the powers that be are resisting?!"

Once again, I persisted and was able to convince the administration to "make an exception," and to secure the funds for health insurance, and allow the Chinese position to go forward as planned. I remember asking myself at the time why we should have to go through these hurdles. After all, I was not an Asian scholar with a personal axe to grind. I was the IEP director, a German professor, trying to help the university be more responsive to the needs of the businesses in Rhode Island, which had thrown in more than its share of resources to this project. Why, I would ask myself, does the institutional mindset so often undermine rather than aggressively find the way to make good things happen? Why probe for the dark side and be so fearful of something new or different?

I have sometimes asked myself if I overstepped my bounds by pushing forward at such a rapid pace with the Chinese program. But I reject that thought when considering the alternative. I could, for example, have taken the idea of a Chinese program to the Department of Languages by means of a department meeting. The result would have been either outright rejection or the creation of a committee to study the feasibility of introducing Chinese at URI. The committee would have been tied up by a variety of doubts and uncertainties, mostly related to finances and demands that the university first fund several new positions in French, Spanish, and Italian. This debate would have gone on indefinitely and most certainly would have been rejected by the department. Rather than take that route, I listened to the business leaders who believed so strongly in the IEP, secured some start-up monies from them, received encouragement and a salaried position from the Chinese government, then helped to bring in a $500,000 donation from the Hasbro Corporation, and, finally, competed successfully for a $1,000,000 grant from the Department of Defense.

Despite the troubles along the way, URI and the IEP have developed a full degree program in Mandarin Chinese and an astonishingly strong Chinese IEP for its young age. There are now two tenure-track faculty, one being Dr. Wen Xiong, who came to us originally through the Chinese government grant and was converted to a regular URI tenure-track position in 2009 after her three years of Chinese support. The other is Dr. Wayne Wenchao He, who came to us as an Associate Professor from West Point with initial support from the National Security Education Program. There are also two lectureship positions which are needed to meet the instructional demand of over 200 URI students enrolled in Chinese classes each semester, with 55 majors. Dr. Xiong wears three hats as the Associate Director of the Chinese IEP and the Chinese Flagship and as director of the Chinese Summer School while Dr. He has an enormous responsibility as director of the Chinese Flagship Program, director of the Confucius Institute, and Associate Professor of Chinese. The funding structure of the Chinese effort at URI requires balancing multiple programs and initiatives at the same time.

The IEP is proud to have become a member of the Chinese Language Flagship Program in 2008 by means of a $1,000,000 grant from NSEP. The Flagship is an extremely demanding and rigorous language program which expects participating students to reach superior (almost native) proficiency by the time of graduation. To expect engineering students to participate in this curricular option for Chinese at URI is almost unrealistic, but we nevertheless have a small number of bright and highly motivated students who choose to join this program each year. Others select the "regular" Chinese major in preparation for their year as an exchange student and professional intern in China. The latter generally attain an advanced level of proficiency which enables them to function smoothly in Chinese, even though not fully reaching the superior level of the Flagship program. At the moment there are 28 engineering students enrolled in the Chinese IEP.

In support of Chinese learning at URI, and to provide sufficient intensive learning for Flagship students before they do their year abroad, URI has developed both winter (semester break) and summer immersion programs on our own campus, and has also opened up the possibility of attendance at Chinese immersion programs at Indiana University and/or one of several locations in China. Summer immersion programs are mandatory for Flagship students starting Chinese at URI. The URI summer program includes a four-week residential immersion program at our own campus followed by four weeks in China and had 36 registrants in 2012.

Much of the organizational and educational progress of the Chinese program is thanks to the hard work and dedication of our faculty and staff. Dr. Wen Xiong came to URI with years of teaching experience in China and a Ph.D. from Monash University in Australia. She adapted fast to the American system and has been very effective in developing the initial curriculum and the specialized courses for our winter and summer immersion programs. Dr. Wayne Wenchao He was also the right choice for the right time, offering his well-established pedagogical and research expertise to the very demanding requirements of the Chinese Flagship Program and the Chinese IEP. Wayne was also tapped to take over the leadership of the URI Confucius Institute, which had a rough start at URI and needed to be more closely integrated with the overall URI program in Chinese. Much of the daily work of building the Chinese program is thanks to Erin Papa, who was hired in the fall of 2008 as the Chinese Flagship Coordinator. Erin has tirelessly promoted

Figure 10.3: Meeting with potential exchange students.

the program, recruited students, created multiple opportunities for them at URI and in China, and nurtured their development. Erin was herself an IEP graduate, having earned her degrees in German and Civil Engineering. After working for a period as an engineer, she soon learned that her heart was more with languages and international education. She thus worked in China as an English instructor and a tutor for our first students sent to our partner university in Hangzhou and became quite proficient in Mandarin herself, all of which prepared her well for her work as the Chinese Flagship Coordinator. For me it is particularly rewarding to see one of our IEP grads now assuming a leadership position in the field of international engineering education. Once she completes her doctorate in education, which she is now doing at URI, she will be in a fine position to advance in this important field. When I consider all of the above, I think my somewhat unorthodox path to an energetic program in Chinese was well worth the pain endured.

CHAPTER 11

Staying Involved after Retirement

Since retiring from my role as director of the IEP, I have used a good portion of my time to build a closer relationship with alumni of the program. I knew them well as undergrads and did not want to lose that contact. I also wanted to see where their lives took them, and what role the IEP education may have played in helping them get started and make progress with their careers. I have done this by staying in touch through Facebook and LinkedIn, but also through alumni meetings organized by the IEP in Rhode Island at the time of my retirement and since then in Germany. The retirement party in 2010 attracted 250 alums, spouses, and friends to a luncheon and over 200 to an evening dinner. In May 2009 and 2012, the IEP meetings in Braunschweig attracted over 100 persons on each occasion, and led to the founding of two German chapters of the URI Alumni Association. Thanks to the close relationship with so many alumni, I was able to interview fifteen of our grads in depth who have been in the workplace for some time, to gain their perspectives regarding what skills they acquired through their language learning and their time abroad and what difference the IEP has made for them in their lives and careers. The results of this study are published in my book entitled: *Going the Extra Mile: University of Rhode Island Engineers in the Global Workplace*.[1]

I learned from the alumni that they had very few regrets, and had made great progress as students in acquiring their new language, in integrating themselves into a new culture, and gaining sensitivity and expertise in cross-cultural communication. They also had gained an understanding of how engineering is taught at a technical university outside the U.S., and how it is practiced in a different cultural setting, affected by differing value sets, approaches, and practices. What surprised me in this interview process was the extent to which the alumni each reported that they had acquired better day-to-day problem solving skills as a result of being on their own in a foreign environment well outside their comfort zones, and how this, in turn, had boosted their belief in themselves and their self-confidence. Adjusting to a different university system, a different dormitory system, a different diet, a different banking system, different traffic patterns, different attitudes, perspectives, and values—all in another language—meant new challenges every day and new problems to be solved, sometimes small, but sometimes major. They each felt that the experience abroad in a non-hand-holding program, involving both study and work, had enabled them to gain confidence and also to raise the personal bar, and to reach and personal levels not before thought possible. We thus

---

[1]Grandin, John, *Going the Extra Mile: University of Rhode Island Engineers in the Global Workplace*, Rockland Press Rhode Island, 2011

Figure 11.1: Speakers at my retirement party: Ana Franco (Siemens); Eric Sargent (BMW); Deirdre Crowley (Dow); Rick D'Ambrosca (entrepreneur in Prague).

Figure 11.2: IEP grads gather for a 2011 meeting in Munich.

have alumni today in leadership positions with companies such as BMW, Dow Chemical, Covidien, and Siemens. Others have started their own companies, gone on to the most competitive graduate programs, and others have taken even more unforeseen routes. Sareh Rajaee, for example, completed both her MD from Brown University and her Masters in Public Health from Harvard University in 2011, which was followed by an internship in vascular surgery at Yale. Like many other IEPers in the field, Sareh believes she would not be where she is today without the IEP experience. In her words: *The IEP experience, especially my year abroad, helped me build confidence in my interpersonal communication skills, in my independence, and in myself as an individual. The IEP showed me what I am capable of, and I am now a stronger, happier, and more independent person because of it.*

Figure 11.3: Sareh Rajaee, MD.

CHAPTER 12

The Broader Message for Higher Education

As a result of my years of experience with the IEP students and the feedback from alumni, I have come to view the undergraduate IEP (BA/BS) program as a successful model which may and should be widely replicated by other institutions and other fields or combinations of fields within our own university structure. If German and engineering make sense in preparing our students for careers in today's workplace, then why not nursing and Spanish, French and pharmacy, Chinese and business, journalism and Japanese, textile merchandising and Italian, or any language with political science or public affairs, and so on? Certainly, the IEP stands as a model for those who would internationalize their university curricula. But, I would even argue that such dual majors with a focus on global education, personal growth, stronger communication skills and self-reliance could help us to redefine what it means to be a liberally educated person in the twenty-first century. Just as the Morrill Act of 1862 sought to adapt higher education for the age of industrialization by combining the agricultural and mechanic arts with the traditional liberal arts, the IEP advocates a rigorous attention to the STEM disciplines in combination with the study of language and culture and extensive experience abroad. We desperately need engineers and other professionals who are professionally savvy in their disciplines, but also educationally rounded as citizens of the world, knowledgeable and informed, precise in their thinking, and confident and capable as communicators. The IEP, therefore, presents itself as a path for merging two sides of the academy, each as important as the other. As higher education struggles in search of solutions to its crises of cost and relevance and wonders what best to offer its students and how, the dual model of science, technology, and the liberal arts, in close partnership with the public and private sectors, deserves close consideration.

The IEP model also has a strong message for the role of the humanities in higher education today, which is a much-debated subject. At a time such as ours, when many students are graduating from college and finding no work or work which makes little use of their education, both students and parents are beginning to doubt the value of a college education, or are willing to pay only for degree programs which promise jobs. One result is dwindling enrollments in traditional liberal arts disciplines. Even at Yale University the numbers of students majoring in English has declined dramatically, as it has in fields like languages, philosophy, or classics.[1]

Rather than put on the blinders or hope for better times with more job opportunities, it makes sense for humanities faculty and university administrators to begin thinking of new ways

[1] See: Yale Daily News, Antonia Woodford, April 18, 2012, Humanities Face Identity Crisis.

to reinvigorate the humanities which are so critical for the acquisition of desperately needed skills: critical thinking, analytical reasoning, careful reading, strong written and oral communication skills, historical and cultural appreciation, understanding of other cultural perspectives, and so on.

The IEP points to the role that the study of language and culture can and must play in the lives and careers of students in professional fields such as engineering. Combining them necessitates give and take on the part of faculties in two major areas, and it requires a restructuring of a field such as language to make it applicable to students outside of the traditional humanities areas. In our case, it has meant that language study would no longer be reserved exclusively for those pursuing literary study, but would be broadened for a wide range of students. It has required us to get over the idea that the humanities are not intended for practical value and to discover that our fields are desperately needed in the global workplace.

It is encouraging to note that other humanities disciplines have likewise begun to raise the question of their role in higher education for the twenty-first century. Robert Scholes,[2] for example, has written extensively about the typical English department curriculum and lamented its exclusive adherence to the study of literary works belonging to a well-defined and therefore restrictive canon. Scholes would have English departments acknowledge their primary role as teachers of reading and writing based on a much wider range of great texts, including such works as the Bible and the Declaration of Independence. Scholes argues that English faculty should not fear making their courses practical and of use to students who want jobs. He notes that the flourishing fields in universities today are those that offer marketable skills, and points out that the humanities do indeed offer and cultivate highly desirable skills. Their stumbling block in the path of change is the faculty who have been fearful to admit that and to promote themselves in such a way.

University administrators should also consider what this means for the overall curriculum of the student majoring in English, philosophy, languages, or any other discipline in the arts or humanities. Just as we expect an engineering student to have a background in the humanities, we could argue conversely that the humanities student should become well versed in issues of modern science and technology. This could well mean a greater role for science, pharmacy, business, and engineering in general education programs, and it could mean minors, second majors, certificate, or dual degree programs in the science and professional fields for their non-majors.

Another major issue encountered along the way is study abroad. Who should study abroad and for how long, and in what language? Hermann Viets and I knew very well in 1987 that the world was shrinking and that it would be very important for future leaders of business and industry to have experience abroad as part of their undergraduate education. We also both knew that very few engineers were learning languages other than English and that study and work abroad were most commonly experiences reserved for humanities and arts majors, and especially young women. Though the IEP has countered these trends by sending a sizeable number of engineering students abroad each year for a full twelve-month experience, and though the awareness has grown nationally to the extent that many engineering colleges across the country are building programs to help their

---

[2]See: Scholes, Robert, English after the Fall: From Literature to Textuality, University of Iowa Press, 2011.

students study abroad, recent statistics are not encouraging. The Open Doors 2012,[3] survey from the Institute of International Education (IIE) reports that of the almost 274,000 American students participating in some form of study abroad this past academic year, just 3.5% were engineers. It also reports that only 0.1% of these students spent the entire year abroad. The national tendency is for greater numbers abroad, yet for shorter periods of time. Thus, there is much work to be done in internationalizing American engineering education.

---

[3] Open Doors 2012: Report on International Education Exchange, Institute for International Education.

CHAPTER 13

Conclusions

This brief account of my travels with URI's International Engineering Program offers some significant lessons about the challenges and fragility of international programs in the United States and the difficulties and challenges associated with entrepreneurial and interdisciplinary activities within the walls of traditional universities. Though every institution has its own character and its own set of players and therewith its own potential programs and associated challenges, there are nevertheless common experiences, guidelines, and principles that may be considered. I offer the following lessons learned:

1. For any such program to succeed, there must be at least one person passionately committed to the concept who has the ability, credibility, energy, and experience to function as the driving force over the long-term. In my own case, I was fully prepared to make a professional jump from Kafka scholar to IEP director, and I was fully prepared to put unlimited amounts of time into the success of this idea. By that point I had also gained and was gaining substantial administrative experience, serving a number of years as head of the URI German program, five years as Associate and Acting Dean of the College of Arts and Sciences, and six years as Chair of the Department of Languages, and I was a full professor. A program like the IEP will not evolve based on good ideas and short bursts of energy. It requires a driving force, sixty-hour work weeks, few vacations, and a commitment for the long run.

2. If any school is serious about internationalizing engineering education, there must be dedicated, passionate, and credible people involved at the leadership level, and ideally, but not necessarily, with the support and flexibility of the institution. The fact that I was already a full professor at that time simplified the process for me. It is an unfortunate fact that my path would not have been open to a younger faculty member, whatever his or her field might be. Building a program such as this would have been too great a distraction and diversion from traditional research obligations and thus a danger to professional security. We can hope for greater flexibility for faculty achievements in the future and a professional path for pedagogy and program development, as well as greater support for cross-disciplinary work, but that day has not yet arrived.

3. Passion and commitment, however, are of little value without clarity of vision as well as the organizational skills to decide where best to place one's energy. The IEP, i.e., a five-year program leading to degrees in both a language and an engineering discipline and featuring internships and study abroad, is a straight-forward and simple idea, to which one can bring

energy, commitment, passion, an entrepreneurial spirit, persistence, and organizational talent. But, as has been shown in this volume, there are many pieces to the pie, and it requires organizational skill to know what to do when and how well to do it. As IEP director, I often compared myself to the circus performer balancing spinning pie plates on a collection of broomsticks. While keeping an eye on every one of them, it is necessary to ensure that none slows down to the point where it will fall. The students, the faculty, the internship program, the grant writing, the housing projects, the fund raising, the relationships with industry, with the advisory board, planning for the next colloquium, and so on—it all has to be done with high energy, but also care and quality.

4. Personality matters. An IEP director must be able to work tactfully and humbly, yet with determination, in order to command respect across the disciplines and cultures. Anger and aggression simply will not get the job done. Because the program brings parties together who do not know each other and puts forward ideas which are often in conflict with the daily patterns and rules of the system, he or she must have patience and be able to persevere in winning over all parties to the goals of the program. Such a leader must often be prepared to move forward, even though there are many throughout the system who will say it cannot be done. I remain convinced that progress can always be made as long as the idea is sound and basically beneficial to all parties concerned. But even when those criteria are fulfilled, one still needs to be braced for the unexpected.

5. Persist if you know it is right, but do so carefully: As I look back over the twenty-five-year history of the IEP and my own involvement, it is clear that the program has relied on my personal commitment, my willingness to work long hours, and my persistence in overcoming the unfriendly sides of academic bureaucracy. As a program that is bound to two different colleges and academic traditions and seeks the allegiances of students and faculty in new and different ways, the IEP has always been destined for challenge. I thus found myself from the very beginning in seeming conflict with people, rules, regulations, and traditions. My personal challenge has always been to find a way to get it done in the face of those who say: "No, that can't be done!"

6. Don't let the system get you down: What I have found to be lacking so often at URI is the will or mindset to make things happen when common sense indicates the action is right. When we should be asking how we can jointly bring about a good thing, there is always someone with a certain degree of authority probing diligently for some reason why we should not be able do it. I have found this to be discouraging on so many occasions, and have had to accept that things will often be more complicated than necessary. The reasons for this vary, sometimes deriving from fear of change, or lack of vision, or a perceived affront to authority, or even jealousy. Whatever the cause, the result for me has been persistence, usually yielding long-term success, but not without risk, anxiety, and, on many occasions, a definite loneliness.

7. Success of any cross-disciplinary effort such as the IEP must yield clear benefits for all sides. In our case, we saw qualitative and quantitative advances for both the engineering and the language sides. Highly qualified and highly motivated students choose URI for the IEP; global companies flock to our support because of the benefits they can accrue; funding agencies and foundations likewise come to our help, because they want to back demonstrably good and proven ideas. Without clear benefits for all parties, success will be short-term, at best.

8. Be alert to opportunities of the moment and of one's own institution and do not let them pass by. Depending upon one's own interpretation, the IEP has succeeded because of a certain amount of good fortune, e.g., the right people at the right place at the right time, but also because of the leadership's ability to recognize the right opportunities and then take advantage of them with rigor and discipline.

9. Do not shy away from taking calculated risks. Looking back over this text, one can see that I took some large leaps of faith or risks throughout the history of the program, when assuming we could build an internship program, when putting together the building projects, when committing to the Chinese program, and so on. Without risk taking, most of these things would simply not have happened.

10. Engaging language faculty is necessary. Language teaching, cultural preparation, organization of the student exchange, and outreach to global companies, to international foundations, and governmental support resources all would have been much more difficult without the help of the language faculty. It is disheartening to see well-meaning, but inexperienced engineering groups try to go the international route without the cooperation of colleagues needed from other disciplines.

11. Outreach, marketing, and recruiting are absolutely essential. Because the idea of combining language and engineering study is far from the minds of high school students, teachers, and guidance counselors, and even parents, the IEP depends on a rigorous program of outreach and the need to get the concept to the attention of potential students before the start of their college years. We might wish that the wisdom of the program were commonplace and obvious to all, but we believe the American public still is years, if not decades away from this stage.

12. Advising, mentoring, and nurturing of students in the program is likewise critical. IEP students have a challenging curriculum and, at age eighteen or so, sometimes find it difficult to focus on the outcome of the program. We, therefore, reach out to them in a variety of ways. One key to our success is our residential learning communities enabling IEP students to live together, to support each other, and to have ready access to the program administration. Regular meetings, newsletters, special guest speakers from industry, panel presentations from alumni, meetings with advanced students and with exchange students from partner schools are likewise important mechanisms for keeping the students focused and on track.

13. Global networking is a key for success. For us this means continual outreach to our partners in industry as well as our partners in foundations and governmental agencies. We depend on them for student internships and eventual job placements, but also for advice and for financial support. The IEP leadership has built relationships with literally hundreds of persons in businesses, universities, and both private and public agencies in North America, Latin America, Europe, and Asia, and is committed to maintaining and fostering these relationships for mutual benefit. Networking means a willingness to travel regularly, but also daily networking in the office by means of telephone and e-mail.

14. In this era of tight budgets and dwindling state resources, international programs must rely on extramural financial support to do their job adequately. An IEP director must, therefore, expect to be a fundraiser and grant writer. The bad news is that this is required, but the good news is that this is a productive area in American higher education today, with clear benefits for those who have the potential to provide financial help. Foundations, funding agencies, corporate leaders, program alums, and private individuals readily understand the importance of preparing young Americans to be savvy in the global marketplace. If a program is solid and doing its job right, it should not be extraordinarily difficult to find financial supporters.

The work is never finished. With all of it successes, its related praise and awards, it has nevertheless been a struggle to define the point at which the IEP can or could declare itself fully institutionalized, with an infrastructure ensuring a long-term future. Much of this lack of security may be attributed to the fact that I as executive director was, according to official organizational charts, not the IEP director funded by a neutral or shared university line, but rather a Professor of German funded by the College of Arts and Sciences. When I retired, it appeared that this problem was solved through the recasting of the position as an IEP executive directorship, to be funded jointly by the deans of the two colleges involved and eventually by the growing endowment. The IEP staff assistant would also be funded by the university by way of the College of Arts and Sciences, as originally intended.

It speaks well of the program that it enjoys the strong and unequivocal support of the deans of the two colleges, Arts and Sciences Dean Winifred Brownell and Engineering Dean Raymond Wright, who agreed to share the main personnel costs for the program as of academic year 2008-2009. Both deans fully understand the complexities of building cross-college programs, but also fully appreciate the benefits of the program for their own constituencies. Of all engineering deans, Dean Wright is the most outspoken champion since the days of Hermann Viets and argues publicly and frequently that the IEP is the number one strength of the College of Engineering, both in terms of attracting high quality students, and in terms of national reputation among peers across the country. With the deans' strong support, URI has hired a new and permanent full-time executive director, Dr. Sigrid Berka. Similar to my own background, Sigrid is also a Germanist by training, having taught as Assistant Professor of German at Barnard College for many years, and has made a similar career change by first coordinating, then leading the MIT-Germany Program for a decade, which is part of the MIT International Science and Technology Initiatives (MISTI) with the goal of globalizing

MIT engineers and scientists by providing hands-on learning experiences (research and internships) in foreign countries, and training them in the language of the country. Sigrid thus brought both her familiarity with foreign language instruction and an extensive network of German company contacts from MIT to the IEP. In addition, URI is also funding unequivocally the full-time staff assistant, now called the IEP Coordinator. All parties believe that this step will mark the end of the growing pains and the full birth of the IEP as a truly cross-disciplinary International Engineering Program.

At the suggestion of URI President Robert Carothers, who was concerned that my retirement could well lead to the loss of much of the IEP's contact network across the globe, I agreed to continue my work part-time for one year leading to my retirement in June 2010, to assist the new IEP leader as needed, and to ensure that she could take full advantage of the scores of companies, foundations, public sector contacts already in place and supportive of the program. That plan worked very well due to the fact that Sigrid Berka and I fully respect and appreciate one another and understand how best to support each other. Sigrid has maintained close contact with me since retirement, and is welcome to make use of my memory of my 23 years of IEP work, as long as it remains in tact!

Has my own journey with the IEP been worthwhile? I can answer that question by saying how fortunate I was to go to work each morning until age 70 with a sense of challenge and excitement. And I can say how thankful I am to be stimulated and excited by the memories that I am now putting down on paper at age 72. But I can answer best by expressing pride in our students, such as those 33 currently abroad, completing their semester of study in Germany, France, Spain, and China, to be followed by their six-month professional internships. At the super-star level, I can point to John Ellwood and Andy Marchesseault who not only completed the undergraduate IEP and then the dual degree masters in cooperation with the Technische Universität Braunschweig, but who were then invited to stay on in Germany to complete their doctorates, or Ahmed Fadl, the first dual doctoral/PhD graduate who now works at NASA. Or I can cite the accomplishments of so many alums, who have secured fine positions with global companies or even gone on to law or medical school. It was an honor to interview fifteen alumni of our program in depth and publish a volume in 2011 about their professional trajectories and their views about the value of the program for their lives and careers.[1]

If I go back to my own beginnings in the field of global education and my initial experience as a twenty-year old in Germany in 1960, I realize that, imperfect as many things are, the goals I set for myself in my early years at URI have been met in so many ways. I could, in a sense, be satisfied if there were just a handful of IEP successes, but there are in fact at least 35 IEPers going abroad every year, not only to Germany, but to Spain, to France, to Mexico, and to China, and over 400 students have completed the program. I hope this account helps illustrate the degree to which persistence, hard work, patience, and tact are required for a program such as the IEP. But I hope it also illustrates that remarkable progress can be made when an idea is sound and beneficial to all parties concerned. And, finally, I hope it illustrates that my journey has been and continues to be extremely rewarding.

---

[1] Grandin, *Going the Extra Mile*.

Bibliography

[1] J.M. Grandin, German and Engineering: An Overdue Alliance, *Die Unterrichtspraxis*, No. 22 (1989), pp. 146–152. DOI: 10.2307/3530187

[2] J.M. Grandin, Deutsch für Ingenieure: Das Rhode Island Programm, in *Das Jahrbuch Deutsch als Fremdsprache*, Vol. 15, (Fall 1989), pp. 297–306.

[3] J.M. Grandin, Language and Engineering: The Next Step, *Proceedings of the Clemson Conference on Language and International Trade*, ed. S. Carl King and Sixto E. Torres (Clemson, South Carolina: Clemson University, 1989), pp. 29–42.

[4] The University of Rhode Island's International Engineering Program, in *Language and Content: Discipline and Content-Based Approaches to Language Study*, ed. Merle Krueger and Frank Ryan (Lexington, Massachusetts: D.C. Heath and Company, 1992), pp. 130–137.

[5] J.M. Grandin and H. Viets, International Experience for Engineers, in *The International Journal of Engineering Education*, Vol. 9, Nr. 1 (1993). pp. 93–94.

[6] J.M. Grandin, Developing Internships in Germany for International Engineering Students, *Die Unterrichtspraxis*, No. 2 (1991), pp. 209–214. DOI: 10.2307/3531030

[7] J.M. Grandin, Educating American Engineers for the Global Workplace, in *Proceedings of the Fourth World Conference on Engineering Education*, October 15–20, 1995, Saint Paul, Minnesota, edited by E. R. Krueger and F.A. Kulacki, University of Minnesota, vol 4, pp. 29–33.

[8] J.M. Grandin and Thomas J. Kim, The International Engineering Program at the University of Rhode Island, in *Navigating the New Engineering World: Proceedings of the 11th Conference on Engineering Education*, July 5–7, 1998, Dalhousie University, Halifax, Nova Scotia, pp. 519–523.

[9] J.M. Grandin and Kristen Verducchi, The International Engineering Internship Program at the University of Rhode Island, in *The Journal of Chemical Engineering Education*, May 1996, pp. 126–129.

[10] J.M. Grandin and Eric W. Dehmel, Educating Engineers for the Global Workplace: A Study of Cross-Cultural Issues, *Journal of Language for International Business*, Vol. 8, Nr. 2 (1997), pp. 1–15.

[11] J.M. Grandin and Doris Kirchner, German and Engineering - ein interdisziplinäres Programm an der University of Rhode Island, in *Wirtschaftsdeutsch International: Zeitschrift für sprachliche und interkulturelle Wirtschaftskommunikation*, WDi 1/99, pp. 109–119.

[12] J.M. Grandin, German and Engineering at the University of Rhode Island: Preparing Students for the Global Workplace, with Jennifer Dail, in *Lernwelten: Eine Zeitschrift des Goethe-Instituts für Deutschlehrende in den USA*, Heft 3, January-August 2000, pp. 9–10.

[13] J.M. Grandin, Globalization and Its Impact on the Profession, in *Realizing Our Vision of Languages for All*, ed. Audrey Heining-Boynton, American Council on the Teaching of Foreign Languages, 2005.

[14] J.M. Grandin, Preparing Engineers for the Global Workplace: The University of Rhode Island, in *Proceedings of the Annual Meeting of the American Society for Engineering Education*, June 2006.

[15] J.M. Grandin, Preparing Engineers for the Global Workplace: The University of Rhode Island," in the *Online Journal for Global Engineering Education*, Vol. 1, Issue 1, Fall 2006.

[16] J.M. Grandin, International Dual Degrees at the Graduate Levels: The University of Rhode Island and the Technische Universität Braunschweig, *Proceedings of the ASEE Annual Conference*, Honolulu, Hawaii, June 2007.

[17] J.M. Grandin, International Dual Degrees at the Graduate Levels: The University of Rhode Island and the Technische Universität Braunschweig, in the *Online Journal for Global Engineering Education*, Vol. 3, Issue 1, Fall 2008.

[18] J.M. Grandin, Why Learn Another Language if the Whole World Speaks English? in *English as the Global Language: Perspectives and Implications*, ed. S. Rajagopalan, the Icfai University Press, 2007.

[19] J.M. Grandin and E.D. Hirleman, Educating Engineers as Global Citizens: A Call for Action, Report of the National Summit Meeting on the Globalization of Engineering Education, March, 2009. In print or accessed at `http://digitalcommons.uri.edu/ojgee/vol4/iss1/`

[20] J.M.Grandin and N. Hedderich, Intercultural Competence in Engineering, in Deardorff, Darla, *The Sage Handbook of Intercultural Competence*, Sage, 2009, pp. 362–373.

[21] J.M. Grandin, Bridging two Worlds, in Downey, Gary and Beddoes, Kacey, Editors, *What is Global Engineering Education For? The Making of International Educators*, Morgan & Claypool, 2011. DOI: 10.2200/S00302ED1V01Y201010GES001

[22] J.M. Grandin, *Going the Extra Mile: University of Rhode Island Engineers in the Global Workplace*, Rockland Press Rhode Island, 2011.

[23] S. Berka, Institutional Strategies: Seven Cases. *Online Journal for Global Engineering Education*, special volume on Educating Engineers as Global Citizens: A Call for Action/ A Report of the National Summit Meeting on the globalization of Engineering Education, edited by John M. Grandin and E. Dan Hirleman, 4, 1, 22, 2009.

[24] S. Berka, The University of Rhode Island Graduate Dual Degree Program with the Technical University of Braunschweig – Its Added Value, Synergies, and Gains for Engineering Students, *Online Journal for Global Engineering Education*, Vol. 6, Issue 1, Article 5. 2011.

[25] S. Berka, Retaining Engineering Students through a January Term German Immersion Study Tour, *Global Business Languages*, Vol. 16, Article 7, 2011.

[26] K. De Bruin, L. O. Erickson, J. Hammadou Sullivan, A Meaningful French Education: Experiential Learning in French, *The French Review*, forthcoming.

[27] L. O. Erickson, I Can Intern in France! Student Perceptions of Success during Their International Engineering Internship, *Online Journal for Global Engineering Education*, Vol. 6: Issue 1, Article 6, 2011.

[28] L. O. Erickson, Blueprint of Technical Professions: Changing Conceptions of Work and Education in Eighteenth Century France, *The French Review*, 85, 6, 2012.

[29] W.W. He, Effects of Short Term Culture and Language Immersion Porgram on Language Learning: A Case Study, Journal of Chinese Language Teachers Association, 43, 3, pp. 65–80, 2010.

[30] W.W. He and Jiao, D., Curriculum Design and Special Features of "Computer Chinese" and Chinese for Tomorrow, in *Teaching and Learning Chinese: Issues and Perspectives*, Charlotte, NC, Information Age Publishing, pp. 217–236, 2010.

[31] N. Hedderich, The Challenge of Transcultural Competence: Background Reading of of Target Culture Current Events Articles, *Global Business Languages*, Vol. 16, Article 9, 2011.

[32] D. Rarick, The Student Centered Classroom Made Real: Transforming Student Presentations in an Advanced Course on Technical German, *Die Unterrichtspraxis/Teaching German*, 43, 1, 2010, pp. 61–69.

[33] W. von Reinhart, German for Science and Technology: Teaching Strategies for Beginning Students, *Die Unterrichtspraxis/Teaching German*, 34, 2, 2001, pp. 119–32.

[34] W. Xiong, Peiyang Quanqiu hua de zhuangye rencai: jiyu he tiaozhan (*Chinese Language Flagship Partner Program for Global Professionals: Opportunities and Challenges*). *Journal of Chinese Teaching and Research in the U.S.* vol. 1, pp. 69–71, May 2011

[35] W. Xiong, Zhongwen zuowei guojihua gongzuo yuyan de sijian he sikao (*Practice and Reflections of Chinese as an International Work Language*). In Li, JH (Eds). *Teaching Chinese in a Global Context*. Chapter 12. Xuelin Publishing. Shanghai, China., pp. 181–193, Feb 2011

[36] W. Xiong, & J. M. Grandin, The Role of Chinese Culture and Language in Global Education: The Chinese International Engineering Program at University of Rhode Island. In Chen, J., Wang, C., & Cai, J. (Eds.). *Teaching and learning Chinese: Multiple perspectives*, Charlotte, NC: Information Age Publishing, 2010.

Author's Biography

JOHN M. GRANDIN

John M. Grandin is Professor Emeritus of German and Director Emeritus of the International Engineering Program at the University of Rhode Island, an interdisciplinary curriculum, through which students complete simultaneous degrees (BA and BS) in German, French, Spanish, or Chinese, and in an engineering discipline. Grandin has received numerous awards for his work combining languages and engineering, including the Federal Cross of Honor (First Class) from the Federal Republic of Germany, the Award for Educational Innovation from ABET, and the Michael P. Malone Award for Excellence in International Education from NASULGC, the National Association of State Universities and Land Grant Colleges, the American Association of Teachers of German (AATG) Outstanding Educator Award, the DAAD Alumni Association Award for International Exchange, and the Association of Departments of Foreign Languages (ADFL) Award for Distinguished Service in the Profession. He has published widely on such cross-disciplinary initiatives and has been the principle investigator for several funded projects related to the development of the International Engineering Program. Grandin also founded and organized the Annual Colloquium on International Engineering Education, bringing together university faculty and business representatives each year to promote a more global engineering education nationally (http://uri.edu/iep.)

The author's son, grandchildren, and IEP students plant a tree at the Heidi Kirk Duffy Center in his honor—-May 2010.

Printed in the United States
by Baker & Taylor Publisher Services